FOOTPRINTS IN THE JUNGLE

FOOTPRINTS IN THE JUNGLE

Natural Resource Industries, Infrastructure, and Biodiversity Conservation

EDITED BY

Ian A. Bowles

Glenn T. Prickett

ASSISTANT EDITOR

Amy E. Skoczlas

OXFORD
UNIVERSITY PRESS

2001

OXFORD
UNIVERSITY PRESS

Oxford   New York
Athens   Auckland   Bangkok   Bogotá   Buenos Aires   Calcutta
Cape Town   Chennai   Dar es Salaam   Delhi   Florence   Hong Kong   Istanbul
Karachi   Kuala Lumpur   Madrid   Melbourne   Mexico City   Mumbai
Nairobi   Paris   São Paulo   Shanghai   Singapore   Taipei   Tokyo   Toronto   Warsaw

and associated companies in
Berlin   Ibadan

Copyright © 2001 by Oxford University Press, Inc.

Published by Oxford University Press, Inc.
198 Madison Avenue, New York, New York 10016

Oxford is a registered trademark of Oxford University Press.

Library of Congress Cataloging-in-Publication Data
Footprints in the jungle : natural resource industries,
   infrastructure, and biodiversity conservation / edited by
Ian A. Bowles, Glenn T. Prickett ;
   editorial assistance by Amy E. Skoczlas.
      p.   cm.
   Includes bibliographical references and index.
   ISBN 0-19-512578-9
   1. Forest management—Tropics.   2. Forest products industry—
Environmental aspects—Tropics.   3. Economic development—
Environmental aspects—Tropics.   4. Biological diversity
conservation—Tropics.   5. Rain forest conservation.   6. Rain forest
ecology.   I. Bowles, Ian A.   II. Prickett, Glenn T.
SD247.F66   2000
333.95′16′0913—dc21   99-18433

The views in this volume are not intended to reflect the views of the Council on
Environmental Equality, the National Security Council, or the U.S. government.

9   8   7   6   5   4   3

Printed in the United States of America
on recycled, acid-free paper

Preface

A sharp increase in investment by resource industries is changing the course of history for delicate tropical ecosystems and the communities who inhabit them. This volume is intended to help all concerned parties—activists, corporations, local communities, governments, and conservation organizations—address this growing challenge.

In the pages that follow, contributors examine the environmental and social impacts of resource extraction and infrastructure development and highlight approaches taken to date to address these issues by both companies and nongovernmental organizations (NGOs). The book is intended to stimulate debate about the feasibility of, and constraints to, developing natural resources in a manner that safeguards biodiversity and respects the interests of local communities. It seeks to highlight emerging "best practices" and focus attention on challenging technical, environmental, social, and legal issues.

The book is organized into five parts. Part I provides the reader with an overview of the book's main topics and themes. In the opening essay, conservationists Russell Mittermeier and William Konstant explain why we should care about biodiversity in the first place. They explain that scientists and the general public understand very little about "biodiversity"—the term used to encompass the wide range of species, ecosystems, and ecological processes that are found on the planet Earth. The authors discuss challenges to conserving this biodiversity and approaches to setting priorities for conservation investments.

Chapter 2 helps set the stage for future chapters by examining new trends in development finance. Everett Santos, of Emerging Market

Partners, formerly a senior official at the World Bank's private sector arm, the International Finance Corporation, describes infrastructure financing as an example of the rapid growth in private sector investment in developing countries and the emerging shift from public to private finance.

Chapter 3 provides a different perspective on the role of extractive industries in environmental protection and economic development. Chris Chamberlain and Lisa Jordan of the Bank Information Center, an NGO that promotes reform and transparency at the multilateral development banks, summarize the range of approaches undertaken by environmental organizations to bring about greater public accountability in both leading corporations and public institutions.

Parts II through V address the specific sectors of oil and gas, timber, mining, and infrastructure. Each part begins with an overview of the key environmental and social challenges in the sector and a summary of emerging best practices. Case studies on a wide range of operations throughout the tropics highlight key issues in depth, including the following: technologies and practices for environmental management; creation of new protected areas in the context of resource development; managing impacts on local communities and supporting community development; independent monitoring and evaluation of environmental and social impacts; product certification and consumer awareness; and legal mechanisms to provide for benefit-sharing among national governments and local communities.

Part II concerns oil and gas development. Chapter 4, an overview by Amy Rosenfeld and her colleagues at Conservation International, describes investment trends, potential impacts, and emerging best practices for petroleum development in the tropics. In chapter 5, Jennifer Parnell and Robert Kratsas, environmental experts with the Atlantic Richfield Corporation, now merged with BP, provide their industry perspective on the same challenges, using case studies to illustrate their points. Chapter 6, by authors from Conservation International and Mobil, presents their experience in ecological monitoring in the exploration stage of oil development in the Peruvian Amazon—the site of a recent surge in petroleum extraction activity.

Part III addresses the onslaught of timber development in tropical forests, painting a challenging picture for conservationists because, unlike oil, gas, and mineral development, logging has a much larger "footprint" and can have far more extensive impacts on biodiversity. Chapter 7, an overview by Nigel Sizer, who leads the World Resources Institute's Forest Frontiers Initiative, summarizes the current state of the world's tropical forests and documents numerous examples of how Pacific Rim nations in particular are driving the expansion of logging into the world's remaining tropical frontiers. In chapter 8, Erling Lorentzen, a leading Brazilian industrialist and chairman of Aracruz Celulose, presents the case for plantations' ability to meet global wood demand. The case study of Aracruz describes how plantations can be sited on degraded lands and be operated in a profitable and environmentally sensitive manner.

In chapter 9 Justin Ward of Conservation International and Yurij Bihun of the Natural Resources Defense Council and Shelterwood Systems International, respectively, present one response to the expansion of logging into tropical forest frontiers: the use of "timber certification" and community forest management techniques in the tropical forests of southern Mexico.

Finally, Part III concludes with a thought-provoking chapter by natural resource economists Richard Rice and Cheri Sugal, and biologists Peter Frumhoff, Liz Losos, and Ted Gullison on strategies for achieving conservation results in the context of tropical forests being subject to logging. The authors provide a powerful counterpoint to those who are advocating "sustainable management" of timber resources in the tropics by pointing out that there are often more direct ways to conserve biodiversity in the context of logging—such as use of primary forest "set-asides"—than through investments in silvicultural "best practices."

Part IV addresses mineral development in the tropics. In chapter 11, professor Alyson Warhurst and her colleague Kevin Franklin at the Warwick Business School at the University of Warwick, England, provide an overview of recent trends and emerging technological improvements in mineral development. In chapter 12, Gary Nash of the International Council on Metals and the Environment (ICME), a consortium of major nonferrous mining companies, provides an industry perspective on the same issues. Nash provides an interesting summary of the use of voluntary guidelines and describes ICME's efforts to educate its members through a variety of case studies.

In chapter 13, Fred Graybeal of Asarco, one of the largest base metal mining companies in the world, summarizes his company's experience with best practices for mining in tropical forests; he highlights Asarco's experience with the creation of an independent Environmental Advisory Committee for a gold mining project in northern French Guiana—a case study that frankly discusses the costs and benefits of this approach. Chapter 14, by David Smith, Vice Dean of Harvard Law School, and Cyril Kormos, of Conservation International, provides an historical perspective on the use of legislation and contracts to safeguard the environment in the context of mineral and petroleum projects in developing countries. They conclude that, while the time-consuming and costly process of developing effective legislation and regulatory capacity unfolds, contracts, insurance, and other legal protections can and should be built directly into the agreements that govern individual projects.

Part V addresses the common thread in major natural resource extraction projects: the infrastructure that ties them to markets and sources of labor, goods, and services. In chapter 15, an overview on infrastructure and challenges for conservation, leading Brazilian industrialist and former Minister of Strategic Affairs Eliezer Batista and his colleagues provide a primer on how development planners can evaluate major infrastructure projects to reduce their ultimate impact on biodiversity. In doing so, this chapter considers the conservation implications both of major "integration" and "penetration" projects. Chapter 16, by Robert Dobias of the Asian Development Bank and Kirk Talbott of Conservation International, considers the specific infrastructure "integration" case in the highly populated greater Mekong subregion in southeast Asia. Dobias and Talbott consider the role of government and of public participation in the many processes that underpin large-scale infrastructure projects.

In chapter 17, John Reid of the Conservation Strategy Fund goes to the other end of the spectrum and considers the conservation implications of new rural roads in more remote areas. In presenting case studies from both the remote Bolivian Amazon and a less developed part of the Brazilian Atlantic coastal forest, Reid summarizes the challenges presented by individual road projects and explains how conservationists can deploy economic cost-benefit analyses to strengthen environmental arguments. Reid also presents the concrete case of how a new park, or conservation "offset," was created in southern Bahia to reduce the impact of road development on biodiversity. Part V concludes with an interesting description of a joint project of the Andean Development Corporation and Conservation International, an innovative new software tool, Cóndor, which is currently being used by infrastructure planners to help avoid impacts on biodiversity.

We introduce and conclude the volume with a broad analysis of trends in natural resource industries and related infrastructure and a set of recommendations for all those concerned with their footprints in the jungle.

Ian A. Bowles
Glenn T. Prickett

Acknowledgments

This book is the product of a great many contributors. First and foremost, the editors are grateful to Ms. Amy Skoczlas for her tireless devotion and leadership as our Assistant Editor. The editors would also like to thank the outstanding group of contributing authors who provided excellent chapters that are the main subject of the book—the authors gave liberally of their time and shared with us and the readers their enthusiasm and experience with complex issues. In addition, the editors would like to thank Robin Bell and Debra Gordon for their critical support in development of the book prospectus and our initial discussions with Oxford University Press. We would also like to thank many Conservation International staff including Lena McDowall, Sterling Zumbrunn, Tanya Tarar, Petra MacGowan, Amy Rosenfeld, Cheri Sugal, Jeff Mastracchio, and Aaron Bruner for their help. Finally, we would like to thank our editor, Joyce Berry at Oxford University Press, for her enthusiasm, encouragement, and flexibility throughout the process that has led to publication of this volume. Ian thanks his family for their encouragement and dedicates this effort to the memory of his grandfather, Dr. George H. A. Clowes Jr., for his scholarly leadership and great love of nature. Glenn would like to thank his wife, Lisa Prickett, for her support and encouragement of this effort, and his father, Gordon Prickett, for his valuable insights as both a mining engineer and environmentalist.

Contents

Part II: Oil and Gas Development Meet Conservation

Part III: Forests under Pressure

Part IV: Mining and Conservation

Part V: Infrastructure for Sustainable Development

Part VI: Conclusion

About the Editors

Until Spring 1999, **Ian A. Bowles** was Vice President of Conservation Policy at Conservation International (CI); he directed CI's work in the areas of policy research, natural resource economics, finance, and law. He is the author of more than twenty published articles on conservation issues; his research at CI focused on the role of development agencies and the private sector in biodiversity conservation. Bowles now serves as Associate Director for International Affairs at the Council on Environmental Quality and as Director of Environmental Affairs at the National Security Council in the White House. This book was written prior to his government service and no views expressed in the book represent a position of the Executive Office of the President or the United States government. He is a graduate of Harvard University and is a native of Woods Hole, Massachusetts, where he is also active on local environmental issues.

Glenn T. Prickett is Vice President for Business and Policy at CI; he leads strategic partnerships between CI and major international corporations in a range of fields, including oil and gas development, mining, and infrastructure development, among others. Prior to joining CI, Prickett served as Chief Environmental Advisor at the U.S. Agency for International Development. He is a graduate of Yale University and resides in Great Falls, Virginia, with his wife, Lisa.

Working in twenty-three countries around the world, CI is a field-based, nonprofit organization that protects the Earth's biologically richest ecosystems and helps the people who live there improve their quality of life. CI seeks to harness the power of the private sector to deliver conservation solutions in the world's most important natural ecosystems. CI's mission is to conserve the Earth's living heritage and global biodiversity and to demonstrate that human societies are able to live harmoniously with nature.

Contributors

Eliezer Batista is the Chairman of Rio Doce International S.A. in Rio de Janeiro.

Yurij Bihun is a consulting forester and Director of Shelterwood Systems International in Burlington, Vermont.

Ian A. Bowles was Vice President of Conservation Policy at Conservation International in Washington, D.C., at the time he edited this book.

Christopher H. Chamberlain is the Project Manager for Central/Eastern Europe at the Bank Information Center in Washington, D.C.

At the time of writing, Robert J. Dobias was the Senior Environmental Specialist in the Office of Environment and Social Development at the Asian Development Bank in Manila.

Joseph Donnaway is Manager of Health and Safety for Mobil Corporation in Dallas, Texas.

Gustavo A.B. da Fonseca is Vice President and Executive Director of the Center for Applied Biodiversity Science at Conservation International.

Kevin Franklin is Research Officer researching biodiversity conservation indicators and minerals development at the Mining and Energy Research Network, Warwick Business School, University of Warwick, England.

Peter C. Frumhoff is Director of the Department of Global Resources at the Union of Concerned Scientists in Cambridge, Massachusetts.

Debra Gordon was formerly Coordinator for Conservation Policy at Conservation International in Washington, D.C.

At the time of writing, Frederick T. Graybeal was Chief Geologist of Asarco Incorporated in New York City.

Marianne Guerin-McManus is Director of Conservation Finance at Conservation International in Washington, D.C.

Raymond Gullison is a Research Associate at the University of British Columbia in Vancouver.

Lisa Jordan is Executive Director of the Bank Information Center in Washington, D.C.

William R. Konstant is Special Projects Director for Conservation International in Washington, D.C.

Cyril Kormos is Director of Conservation Policy at Conservation International in Washington, D.C.

Robert Kratsas was Manager of the Environment, Health, and Safety Department of ARCO International Oil and Gas Company in Plano, Texas.

Erling Lorentzen is Chairman of Aracruz Celulose in Minas Gerais, Brazil.

Elizabeth Losos is Director of the Center for Tropical Forest Science at the Smithsonian Tropical Research Institute in Washington, D.C.

Claudia Martinez is Head of Coordination for Sustainable Development at Corporación Andina de Fomento in Caracas, Venezuela.

Carol Mitchell is EISA Environmental Coordinator for Conservation International in Lima, Peru.

Russell A. Mittermeier is President of Conservation International in Washington, D.C.

Gary Nash is Secretary General of the International Council on Metals and the Environment in Ottawa, Canada.

Silvio Olivieri is Vice President for Strategic Planning at Conservation International in Washington, D.C.

At the time of writing, Jennifer A. Parnell was Senior Environmental Engineer in the Environment, Health, and Safety Department of ARCO International Oil and Gas Company in Plano, Texas.

Richard Piland was Director of Projects for Conservation International in Lima, Peru, at the time he contributed to this book.

Glenn T. Prickett is Vice President for Business and Policy at Conservation International in Washington, D.C.

John Reid is President of the Conservation Strategy Fund of the Pacific Forest Trust in Boonville, California.

Richard Rice is Chief Economist in the Center for Applied Biodiversity Science at Conservation International in Washington, D.C.

Amy B. Rosenfeld is Director of Energy and Mining Industry Initiatives at Conservation International in Washington, D.C.

Everett J. Santos is Chief Executive Officer of the Latin America Group of Emerging Markets Corporation in Washington, D.C.

Nigel Sizer is Team Leader and Senior Associate of the Latin America Group at the World Resources Institute in Washington, D.C.

David N. Smith is the Vice Dean of Harvard Law School in Cambridge, Massachusetts.

Cheri Sugal is Manager of Carbon Offsets and the Tropical Wilderness Protection Fund at Conservation International in Washington, D.C.

Kirk Talbott is Senior Director for the Asia and Pacific Region at Conservation International in Washington, D.C.

Jorgen B. Thomsen is Vice President for Monitoring and Evaluation, and Development Agency Relations at Conservation International in Washington, D.C.

Justin R.Ward is Director of Agriculture and Fisheries Industry Initiatives at Conservation International in Washington, D.C.

Alyson Warhurst is Director of the Mining and Energy Research Network, Warwick Business School, University of Warwick, England.

FOOTPRINTS IN THE JUNGLE

Introduction

*The Growing Footprint: Resource
Extraction Investments Expand
Further into the Humid Tropics*

Ian A. Bowles
Glenn T. Prickett

Extractive industries (logging, mining, oil and gas development) and the infrastructure associated with them (roads, pipelines, transmission lines) have had significant environmental impacts worldwide. Today the potential impact of these industries on global biological diversity is growing, as they focus increasingly on new opportunities in resource-rich developing countries, including some of the world's most sensitive and poorly understood natural ecosystems. In the next decade, for example, it is expected that a significant portion of all new oil exploration and production will occur in the humid tropics. The discovery of new reserves and the liberalization of developing country economies are pressing development further into the last undisturbed tropical ecosystems—home to the world's greatest biodiversity and the least contacted indigenous cultures.

Tropical Latin America is a case in point. Areas like the Amazon regions of Peru and Colombia, as well as northern Guatemala, were, until recently, unattractive for major petroleum development due to political and economic instability. In recent years, however, much has changed. For example, Colombia is now host to the largest oil discovery in the Western Hemisphere in twenty years. Peru is host to a similarly large natural gas discovery at the Camisea field. And in Guatemala, oil development in the Petén region—largely in areas that overlap the ecologically sensitive Maya Biosphere Reserve—has returned after years of being largely stalled. The same trends hold true for mineral development. Between 1991 and 1996, investment in mineral exploration in Latin America more than doubled, from $300 to $700 million. Indeed, a 1996 survey of mining company executives

3

indicated that seven of the top ten countries that were believed to have the best opportunities for investment were in Latin America. The region has also seen a rapid increase in logging investments—timber exports from Latin America jumped 120 percent in the period from 1990 to 1996.

As the number of resource concessions and associated infrastructure projects in tropical countries rises, public concerns about their environmental and social impacts increase as well. Recent experiences in Nigeria, Indonesia, Colombia, Papua New Guinea, Ecuador, and other countries highlight a growing trend of controversy around resource extraction operations. Because developing countries depend on revenue from resource extraction, however, it is clear that development of the world's tropical ecosystems will continue in spite of the controversies.

Choices made now will determine the "footprint" of these industries—the extent of disturbance to natural habitats and the impacts on local communities. The question for all those concerned with the conservation of global biodiversity is how best to respond to this new challenge. We believe that a number of important lessons can be derived from the experiences summarized in this book.

First, resource companies can take steps to dramatically reduce their impacts on ecosystems and communities seen in earlier projects. The case studies in this book describe emerging "best practices" for minimizing direct disturbance of habitats, preventing illegal colonization that leads to further destruction, and responding to the interests of local communities. Due to the weakness of regulatory regimes in many developing countries, adoption of these practices will often depend on a company's voluntary willingness to pursue them. Therefore, the policies and motivations of individual companies—and the vigilance of local environmental and community advocates—will be key determinants of a particular project's environmental and social "footprint."

But the evidence in this book also suggests that biodiversity will not be conserved by the use of better management practices alone. While information and improved global communications mean that companies are more readily accountable for their actions in remote locales, the use of better practices and technologies in each footprint can only do so much. In the end, biodiversity will be largely conserved in the natural, pristine areas that are not subject to intensive resource development. This means that conservationists, corporations, local communities, and governments must include investments in conserving natural areas as part of the matrix used to judge the environmental performance of given projects. *Proactive* investment in biodiversity conservation should be considered part of the "cost of doing business." For conservation to be successful, better management practices by themselves will not be enough.

Other important lessons that we derive from the chapters presented are as follows:

- *Investment in new protected areas.* Creation of, and funding for, protected areas should be an integral part of any development planning. This strategy should include increased investment in effective management of existing protected areas and in the establishment of new areas to provide maximum representation of a region's biological wealth. In more remote areas with extensive and mostly intact ecosystems, the time is right for major new investments in conservation on a larger scale.
- *No development in protected areas.* Protected areas are the most critical core element of national strategies to conserve biological diversity, as they are the only areas where plant and animal species that are not able to exist in human-modified landscapes can survive. Natural resource extraction should simply not be permitted in these areas.
- *Minimizing impacts.* Where petroleum and mineral development does occur in sensitive ecosystems like tropical forests, companies and conservationists should engage in an informed debate about approaches to minimizing the social and environmental impact of exploration and extraction activities. Industry associations, for example, can provide an appropriate forum for dissemination of information on such approaches. Conservationists must also inform themselves of the relevant technical information in order to engage with potential investors in an informed dialogue.
- *Limiting the geographic extent of logging.* Whereas technology and better practices can be deployed to significantly reduce the impact of mining, petroleum extraction, and infrastructure development on tropical forests and biodiversity, timber extraction inherently has a much more extensive geographic impact. While much of the tropical forestry debate has centered around promotion of natural forest management as a conservation tool, there are significant economic, ecological, and institutional impediments to its use in tropical forests. Conservationists should perhaps instead work even harder to stay focused on direct questions of logging impacts. Simple requirements like area fee taxes, conservation "set-asides" in individual timber concessions, and even the retirement of lightly logged areas can often do more for biodiversity conservation than complex changes in silvicultural management techniques.

- *Capacity building and local participation.* All major development projects should take place within the context of public participation in decisionmaking. Local communities, municipalities, and other stakeholders should be involved in a public dialogue with companies, governments, and financial institutions at all steps of the development process. This process has not taken place in the past but must be an integral part of development in the future.

- *Regional planning.* International financial institutions, national governments, and business organizations alike are showing new interest in integrating biodiversity conservation considerations into regional planning. Conservationists should cooperate with such efforts to ensure that the next generation of large-scale zoning and regional planning efforts includes explicit efforts to steer roads and investment away from areas of greatest importance for biodiversity conservation.

- *Further use of legal tools and economic analyses.* Conservationists and governments alike can, and should, support broader use of innovative legal and financial tools to strengthen the environmental performance of private operators. For example, concepts like the creation of "performance bonds" can help reduce the risk of environmental noncompliance. Similarly, the use of simple economic analyses can broaden the technical basis for conservation advocates. Many road development projects often rest on shaky economic underpinnings. Timber development, in certain instances, is also predicated on faulty economic performance assumptions.

- *Ecological monitoring.* Monitoring of the ecological impacts of major natural resource extraction projects should grow to become a fundamental part of industrial planning in sensitive ecosystems. Well-developed ecological monitoring protocols can provide an important baseline for assessing environmental performance.

There is, of course, no "one-size-fits-all" answer to the challenge of conservation. We offer this book as a resource to those concerned with finding new approaches appropriate to their own challenges.

PART I

Conservation and Development in
the Twenty-first-Century Tropics

1 Biodiversity Conservation

Global Priorities, Trends, and the
Outlook for the Future

Russell A. Mittermeier
William R. Konstant

Biodiversity, simply stated, is the total expression of life on Earth. The first living organisms, which appeared nearly three billion years ago, were microscopic, unicellular, and, quite literally, a drop in an immense sterile ocean that covered the planet at that time. Through countless millennia of progressive and sometimes catastrophic evolutionary processes, these first fragile experiments with life have yielded the diversity we know today—an enormous yet relatively thin mantle of microbes, fungi, plants, and animals that covers the Earth, a myriad of species of which ours is but one.

We are fast coming to realize that the condition and survival of the human species will ultimately depend on our ability to maintain existing levels of biodiversity and essential ecological processes. Our planet currently faces an array of environmental problems; some we have been dealing with for centuries, others have emerged more recently as global threats. While some progress has been and continues to be made with regard to air and water pollution, hazardous waste disposal, and soil erosion, we are still desperately trying to fathom the consequences of ozone layer depletion and climate change. Furthermore, we have the immense problem of explosive human population growth, especially in the developing countries, and the excessive resource consumption of the developed nations.

Critical though all of these issues may be, we believe that there is one environmental issue that surpasses all others in terms of long-term global impact: that is loss of our planet's biological diversity, that wealth of species, ecosystems, and ecological processes that makes our living planet what it is. After all is said and done—all the

9

recent space probes searching for life on Mars, the moons of Jupiter, and elsewhere notwithstanding—Earth is still the only place in the entire universe where we know, with certainty, that life exists.

Why is the loss of biodiversity so important? The answers are very straightforward. First, such losses are *irreversible*, at least within any time frame meaningful to our own species. Although humanity already has, or can certainly develop, the technologies to combat most other environmental ills—even if it sometimes lacks the political will or economic incentive to put them into effect—once a species of plant or animal goes extinct, it is gone forever and will never be seen again. And right now, the world faces not just the loss of individual species but the degradation and disappearance of entire biological communities and ecosystems. The second answer is that humans, as living creatures themselves, depend on a vast array of other forms of life for their own survival. Although those in the developed world sometimes lose track of this because they are somewhat insulated from their direct dependence on other life forms by their technology and their sophisticated market systems, that dependence, nonetheless, is still there. For these reasons, we strongly believe that maintenance of the planet's biodiversity is simply the most fundamental of all environmental issues and probably the single most important issue of our time.

Surprisingly, given the incredible advances in science during the present century, our measure of biodiversity remains embryonic at best. Biologists have thus far described approximately 1.4–1.8 million living creatures (Parker 1982). However, estimates and projections made in the last two decades indicate that total species diversity on Earth could be as great as ten million, thirty million, or perhaps even one hundred million or more (Erwin 1983). While we can send members of our own species to the moon and spacecraft to the farthest reaches of the solar system, or fit millions of bits of rapidly retrievable information on tiny computer chips, the truth is that we do not know, probably to within two orders of magnitude, how many species share this planet with us (Wilson 1985; Mittermeier et al. 1997). Needless to say, our ignorance does not just relate to sheer numbers of life forms. If we take into account the endless array of interactions among species, and the ecological processes dependent on and deriving from such interactions, our level of ignorance increases by several additional orders of magnitude—indicating quite clearly that, in many ways, our wonderful technology notwithstanding, we are still in the Dark Ages in terms of our understanding of life on Earth.

Beyond the basic science of biodiversity, we also have much to learn of the economic value of biodiversity to our own species. Recognition over the past few decades of our dependence on biodiversity and the push for a better understanding of the economics of the much-touted concept of "sustainable development" has led, albeit slowly, to increased emphasis on the need for better "valuation of biodiversity." Although as a science such valuation is in its infancy and still lacks appropriate metrics or standardized methodologies, it is finally receiving some attention and we hope will grow significantly in the near future.

The Value of Biodiversity

In relation to the principal theme of this book, we believe that a fundamental understanding of biodiversity value is essential if we are to enter into a meaningful discussion of the importance of different land-use practices, particularly the impact of the wide array of extractive industries. For example, if the biodiversity of a given tract of tropical forest is valued at zero, then even the most uneconomic, unsustainable, and damaging of extractive activities will be seen as generating a net profit for the country or community in question. However, if the true value of watershed protection, the potential or actual export value of certain key forest products, and the range of goods and services that the forest provides to local communities are taken into consideration and assigned real dollar values, then the equation changes dramatically. In the absence of such data on biodiversity use—the prevalent situation in most of the tropical world even today—the result is too often the same: outdated extractive practices carried out with little or no attention to, or concern for, impacts on biodiversity or monitoring thereof.

In an effort to spur such valuation of biodiversity use and to at least categorize levels of use to some extent, one of the authors several years ago defined six distinct categories of biodiversity use (Mittermeier and Bowles 1993). We briefly summarize these here, not because we believe they are the final word but, rather, to stimulate further discussion and interest in a topic that is so fundamentally important to questions of extractive industries and their impact on biodiversity.

Major ecosystem functions. Biodiversity contributes enormously to the regulation of atmospheric chemical composition, temperature, and hydrological cycles, as well as to soil formation, nutrient cycling, pollination, biological control, primary food production, and other major ecosystem functions. A recent study by Robert Costanza and others (1997) evaluated seventeen different ecosystem services (the benefits human populations derive directly or indirectly from ecosystem functions) across the Earth's terrestrial and marine biomes and arrived at a total global value of $33 trillion per year, approximately half of which is attributed to nutrient cycling. Even if this estimate is an order of magnitude too high, the figure is truly staggering.

International trade. The importance of biodiversity to agriculture is one of the more obvious uses of this resource. The annual global value of agricultural trade is in excess of $3 trillion, with basic human nutrition depending largely on seven domesticated species of grass (rice, wheat, barley, oats, sorghum, millet, and corn), all of which require continued genetic input from wild relatives to maintain resistance to diseases and pests. Thus, protecting the diversity of these wild species is critical to sustaining current and future levels of global grain consumption. The situation is similar for coffee, the world's second most traded commodity after petroleum (Myers 1983). To maintain the viability of domestic coffee crops now grown throughout the world's tropical regions, it is critical that wild relatives be conserved in centers of origin such as Ethiopia, the highlands of east Africa, and the rain forests of eastern Madagascar. Madagascar is particularly interesting, since it harbors more than fifty wild species of coffee, several of which are caffeine-free. What is true of these major crop species is also true of the hundreds of other plant and animal species on which we depend for our daily sustenance, and while the global interconnectedness of this dependence is not always obvious, it remains fundamentally important.

Biodiversity also has tremendous implications for industry in the twenty-first century and must be seen as the raw material for the exploding biotechnology industry. Not surprisingly, it is already of considerable importance in industry. Approximately 25 percent of all pharmaceutical prescriptions in the United States contain active ingredients of plant origin, and more than three thousand antibiotics are derived from microorganisms, all percentages and numbers that are likely to increase dramatically in the years to come (WRI, IUCN, and UNEP 1992). In addition, traditional medicine in developing countries, practiced by at least three billion people, relies entirely on the diversity and availability of wild species. The growing herbal remedy market in the developed countries is equally dependent on this source. Industrial uses of biodiversity are not limited to pharmaceutical and medicinal plants. A wide range of wood products, fibers, oils, dyes, resins, latex, tannins, and other products from natural ecosystems have growing commercial use and potential (McNeely et al. 1990).

The enormous recreational uses of biodiversity are well documented. Increasingly, people are looking to the natural world for the maintenance of spiritual and psychological well-being. Nature-based forms of recreation (e.g., fishing, hunting, hiking) are growing in popularity, especially in the industrialized nations. Ecotourism, now estimated at $200 billion annually, is the fastest growing sector of what is now considered the world's largest industry, tourism. Most experts agree that ecotourism will grow most rapidly in the tropical regions, where it is already a major foreign exchange earner in several countries. However, the most appealing flagship species, "charismatic megavertebrates" like the rhinos, the elephants, the larger primates, the big cats, the bears, the savanna ungulates of Africa, the whales, the eagles and other raptors, the birds of paradise, the parrots and macaws, the hornbills, the whales, the marine turtles, and other spectacular creatures, are often the first to go when ecosystem degradation takes place. They must be fully protected if their true potential as sustainable foreign exchange earners is to be realized.

High levels of biodiversity serve to increase the future benefits of scientific research, which in the past was not recognized as a significant earner of foreign exchange but increasingly should be in the future. As interest in biodiversity grows, more and more people will want to carry out research on it, especially in the more remote corners of the tropical world. Moreover, as pristine natural areas continue to disappear, they become an increasingly scarce commodity that will be of ever greater value to those countries with enough foresight to conserve them. A handful of countries, like Costa Rica, already benefit greatly (to the tune of millions of dollars) from scientific research of many different kinds carried out by a wide variety of researchers, ranging from high school students carrying out their first-ever science projects to some of the world's leading lights in conservation biology. Costa Rica's National Biodiversity Institute (INBIO) carries out many different kinds of scientific research, much of it with visiting foreign scientists, and has become a magnet for international funding. This sort of nondestructive biodiversity use is certain to increase significantly in the future and to compete well with the kinds of unsustainable logging activities that unfortunately are still prevalent in many tropical forest areas.

Finally, biodiversity is valuable in international markets as a source of natural products. This includes timber from forest ecosystems; nontimber forest products such as rubber, Brazil nuts, and rattan; animal products like fish, meat, and skins; and a variety of other products that routinely cross international borders. The range of internationally exportable products is also likely to expand significantly in the future, as improved transportation makes a wide range of locally used tropical products (e.g., rain forest fruits) more accessible to international markets.

Regional markets. Despite the international values of biodiversity, the vast majority of its uses will remain within the borders of a particular nation (regional, local, and household values). Products that enter into the regional economy may change hands several times and be transported hundreds of kilometers before they are consumed. For example, products coming into the markets of large Amazonian towns and cities like Manaus, Belém, and Iquitos often originate from a wide radius around them.

Local markets. This category is similar to regional market values, except that the products are harvested and sold in a more circumscribed area and there are probably fewer exchanges between source and market. We also believe that the overall biodiversity value of local markets probably exceeds that at the regional level.

Household use. The issue of household use is often overlooked in valuation of biodiversity. Products do not change hands but are used by the individual who collects or produces them (and his/her family or immediate circle). Examples include fish and game animals for meat, vegetable foods gathered from the surrounding ecosystem, thatch and timber used in building, fuelwood, and medicinal plants, to name just a few. The value of these products is measured not only in market prices but also in terms of their replacement cost should they become unavailable due to environmental degradation or species loss. Since replacement products usually have to be imported from elsewhere, the cost to the individual increases substantially, sometimes to the point that replacement simply does not take place, leading to reduced quality of life.

Global intangibles. Biodiversity is not static but rather is a dynamic, evolutionary complex of ecosystems that function together as environmental buffers to climate change and other global processes. It also has great importance in maintaining worldwide geopolitical stability, sometimes referred to as ecosecurity. Developing countries depend on biological resources on a more direct and immediate basis than do developed nations; the negative impacts of biodiversity loss are confronted on a daily basis as shortages of essentials such as fuelwood, food, and fiber. If not remedied, these crises will force emigration from affected areas, create political instability, and may coalesce into national and regional conflicts. In the Western Hemisphere, these processes are already in evidence in the overpopulated and biologically impoverished countries of Haiti and El Salvador; in the Old World they have been all too obvious in places like West and Central Africa and parts of South Asia.

Understanding the scope and value of biodiversity is essential to dealing with the threats that face it on a global level. Harvard biologist Edward O. Wilson effectively summed up the situation a decade ago:

> Biological diversity must be treated more seriously as a global resource, to be indexed, used, and above all, preserved. Three circumstances conspire to give this matter an unprecedented urgency. First, exploding human populations are degrading the environment at an accelerating rate, especially in tropical countries. Second, science is discovering new uses for biological diversity in ways that can relieve both human suffering and environmental destruction. Third, much of the diversity is being irreversibly lost through extinction caused by the destruction of natural habitats, again especially in the tropics. Overall, we are locked into a race.

We must hurry to acquire the knowledge on which a wise policy of conservation and development can be based for centuries to come. (1988: 3)

While we may still be a long way from documenting the full extent of Earth's living natural heritage, we are becoming painfully aware of the rapid rate at which it is disappearing. Largely through human activities, we have already lost more than 60 percent of the planet's primary tropical rain forest—perhaps the richest of all natural ecosystems—and perhaps a quarter of what remains is degraded (Johnson and Cabarle 1993). Many of the world's coral reefs and other species-rich coastal ecosystems are similarly under threat. According to the *1996 IUCN Red List of Threatened Animals*, fully 25 percent of all mammals and 11 percent of all birds are critically endangered, endangered, or vulnerable, and these figures are surely underestimates (Baille and Groombridge 1996). Furthermore, the *1997 IUCN Red List of Threatened Plants* confirms that nearly thirty-four thousand plant species, or 13.6 percent of the world's vascular flora, are threatened with extinction, again probably an underestimate simply because of lack of information (Walter and Gillett 1997).

Once we confront our ignorance of biodiversity, the fact that it is fast disappearing, and the fact that its loss is a threat to our own survival, we can identify several key issues that need to be addressed in order to do something about the problems that we face and to take meaningful steps to conserve what remains. We certainly believe that a greatly increased emphasis on documenting, cataloguing, and better understanding the full range of biodiversity is fundamental and that it should proceed as rapidly as possible. However, when looking at the great urgency of many of the problems, we recognize that it is not possible to wait for the final results of such research before taking action. Consequently, we believe that two issues are of paramount importance over the short to medium term; we summarize them in the pages that follow. First, priorities should be set to ensure that energy and resources are applied where they will have the greatest impact; second, the critical role of protected areas in achieving biodiversity objectives should be recognized.

Setting Priorities

Conservation International (CI) has been a leader in developing strategies to set conservation priorities in order to maximize the efforts of governments and nongovernmental organizations (NGOs). CI's priority-setting approaches are based on four premises: (1) the biodiversity of every nation is critically important to that nation's survival and must be a fundamental component of any national or regional development strategy; (2) biodiversity is by no means evenly distributed on the Earth's surface, and some areas, especially in the tropics, harbor far greater concentrations of biodiversity than others; (3) some of these areas of highest concentration (e.g., tropical rain forests and coastal ecosystems) are under the most severe threat (already reduced to 25 percent or less of their original extent); and (4) to achieve maximum impact with limited resources, global conservation strategies should concentrate heavily on the areas richest in diversity and most severely threatened, and this investment in conservation should be roughly in proportion to each region's overall contribution to global biodiversity.

Hotspots

The first person to clearly articulate this approach was Norman Myers, one of the leading thinkers of the environmental movement. In a groundbreaking paper, he recognized ten tropical forest "hotspots" that he estimated contained 13.8 percent of all plant species in a mere 0.2 percent of the planet's land area (Myers 1988). In a subsequent analysis, he added several other rain forest areas and four Mediterranean-type ecosystems, increasing the total number of hotspots to eighteen, which then accounted for 20 percent of global plant diversity in just 0.5 percent of Earth's land area (Myers 1990). In 1989, CI (1990a,1990b) and the MacArthur Foundation adopted Myers's hotspots as the guiding principle for their conservation investment, with CI slightly modifying and expanding his list to include areas overlooked in the original analyses (see figure 1.1).

In March 1996, CI began a reanalysis of the hotspots. This analysis was heavily driven by endemism, and especially plant endemism, but also uses data on diversity and endemism of four groups of vertebrates: mammals, birds, reptiles, and amphibians. To qualify as a hot-

---

Figure 1.1 *Biodiversity Hotspots*: 1. Tropical Andes, 2. Mediterranean Basin, 3. Madagascar/Indian Ocean Islands, 4. Mesoamerican Forests, 5. Caribbean Islands, 6. Indo-Burma, 7. Atlantic Forest, 8. Philippines, 9. Cape Floristic Region of South Africa, 10. Eastern Himalayas, 11. Sundaland, 12. Brazilian Cerrado, 13. Southwest Australia, 14. Polynesia and Micronesia, 15. New Caledonia, 16. Chocó/Darien/Western Ecuador, 17. Western Ghats and Sri Lanka, 18. California Floristic Province, 19. Succulent Karoo, 20. New Zealand, 21. Central Chile, 22. Guinean Forests of Western Africa, 23. Caucases, 24. Eastern Arc Mountains/Coastal Forests of Tanzania and Kenya; 25. Wallacea; *Major Tropical Wilderness Areas*: A. Amazon Basin, B. Congo Basin, C. The Island of New Guinea, and Adjacent Archipelago

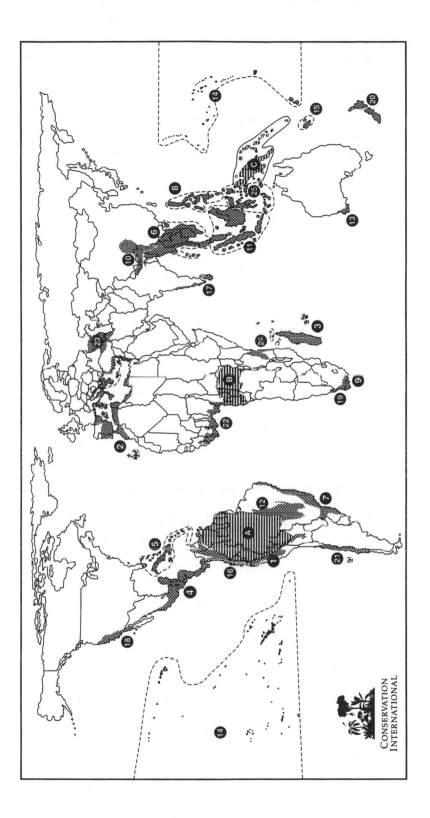

CONSERVATION INTERNATIONAL

spot, an area or region must possess at least 0.5 percent of total global vascular plant diversity. Since current estimates of diversity of vascular plants (angiosperms, gymnosperms, ferns, and relatives) is 250,000, this means that an area must have, *as endemics*, at least 1,250 species living within its borders. It must also have lost at least 75 percent of its original natural vegetation, with only 25 percent or less remaining in more or less pristine condition (in fact many of these areas have already lost 90–98 percent of their original natural vegetation). The reanalysis also considered patterns of diversity and endemism for mammals, birds, reptiles, and amphibians, mainly to complement the plant data and to allow for priorities to be set even within the hotspot list. In general, tropical rain forest areas, which are very rich in plants, also tend to have high levels of diversity and endemism in the four vertebrate groups investigated. In contrast, temperate hotspots, such as the Mediterranean-type ecosystems, and drier tropical hotspots, though exhibiting high levels of plant diversity and endemism, sometimes even exceeding those of the tropical forests, have much lower levels of vertebrate diversity and endemism.

This most recent analysis strongly emphasized endemism for several reasons. First, doing so is in line with the Doctrine of Ultimate Responsibility of the International Union for the Conservation of Nature and Natural Resources (IUCN), which recognizes the great responsibility that a country has if it is the only place where a particular species occurs (McNeely et al. 1990). Second, many endemic species are entirely dependent on a single area for their survival. Third, endemics, due to their restricted ranges and sometimes high degree of specialization, often represent the most vulnerable taxa of a particular community, will almost certainly be the first to fall victims to extinction, and are thus most in need of rapid conservation action (McNeely et al. 1990; Balmford and Long 1994; Pimm et al. 1995). In the current analysis, the number of hotspots increased to twenty-five. Included among these (listed from north to south and west to east) are:

- California Floristic Province: California, Oregon, and Baja California, Mexico;
- Caribbean Islands, including also the southern tip of Florida;
- Mesoamerican Forests, from southern Mexico to the Panama Canal;
- Chocó/Darien/Tumbesian Region, covering the Darien of Panama, Chocó of Colombia, coastal Ecuador, and coastal northern Peru;
- Tropical Andes: Venezuela, Colombia, Ecuador, Peru, and Bolivia;
- Atlantic Forest Region of Brazil, including adjacent portions of Paraguay and Argentina;
- Brazilian Cerrado;

- Central Chile;
- Mediterranean Region of Europe, North Africa, and the Near East;
- Guinean Forests of West Africa;
- Cape Floristic Region of South Africa;
- Succulent Karoo of South Africa and Namibia;
- Eastern Arc Mountains and Coastal Forests of Tanzania and Kenya;
- Madagascar and Indian Ocean Islands;
- Caucases;
- Western Ghats of India and the island of Sri Lanka;
- Eastern Himalayas;
- Indo-Burma Region: northeastern India, Burma, Thailand, and Indochina;
- Philippines;
- Sundaland: Indonesia, Malaysia, Singapore, and Brunei;
- Wallacea: eastern Indonesia;
- Southwestern Australia;
- New Caledonia;
- New Zealand; and
- Polynesia/Micronesia island complex.

Of these hot spots, nine are entirely tropical rain forest areas, five include tropical rain forest and tropical dry forest components, three include tropical rain forest, dry forest, and arid systems, five are temperate Mediterranean-type ecosystems, one is non-Mediterranean temperate forest and shrubland, one is a mosaic of dry forests and savannas, and one is an arid region (Mittermeier et al. 1998).

It is important to point out that this hotspots analysis focused entirely on terrestrial ecosystems. Although some freshwater ecosystems are covered in the regions, we do not claim that these hotspots adequately cover freshwater biodiversity conservation priorities. However, these twenty-five highest priority hotspots, while occupying less than 2 percent of the Earth's land surface, do contain about 45 percent of all vascular plants *as endemics* and between 30 and 40 percent of all nonfish vertebrates *as endemics*. This means that at the very least, they are home to 50 percent of all terrestrial biodiversity (and probably much more) and at least three-quarters of the world's *most endangered* terrestrial biodiversity. The conservation message should ring loud and clear: A very large percentage of global terrestrial biodiversity can be protected in a very small percentage of Earth's real estate.

The next challenge in developing the hotspot strategy is to institute a hierarchy of priority-setting exercises that sharpens the geographic focus (global to regional to national to local to specific sites) or, to put it another way, identifies the "hotspots within the hotspots." For example, the Tropical Andes hotspot, by itself, has twenty thousand plant species *as endemics* (7.4 percent of the global total), has some forty-five thousand plant species in all, and covers over one million square kilometers.

Fortunately, a fair amount of this more specific priority-setting work has already taken place. Among the examples of priority-setting activities within hotspots are CI's Regional Priority-Setting Workshops (e.g., Conservation International et al. 1994, 1996; Ganzhorn et al. 1997), expeditions of CI's Rapid Assessment Program (RAP) (Parker and Bailey 1991; Parker and Carr 1992; Parker et al. 1993a, 1993b; Foster et al. 1994; Schulenberg and Awbrey 1997a, 1997b), development of BirdLife International's Endemic Bird Areas concept (Bibby et al. 1992; Stattersfield et al. 1998), and the analysis of Centres of Plant Diversity of the World Wildlife Fund (WWF) and IUCN (WWF and IUCN 1994, 1995, 1997). Nonetheless, it is clear that much more still needs to be done, and CI alone has completed or is planning several other RAP expeditions in the Tropical Andes, Madagascar, Sundaland, and Wallacea, as well as Regional Priority-Setting Workshops in the Guinean Forests of West Africa, the Tropical Andes, and the Atlantic Forest Region of Brazil, with the Atlantic forest exercise already the third of its kind in this decade.

Major Tropical Wilderness Areas

This priority category deals also with high-biodiversity areas but those at the other end of the threat spectrum from the hot spots. This concept was developed more or less simultaneously by Myers (1988, 1990), who referred to these ecosystems as good news areas, and by Mittermeier (in McNeely et al. 1990), and represents a second focus of biodiversity priority-setting. In contrast to the hotspots, major tropical wilderness areas retain upwards of 75 percent of their original forest cover and also exhibit low human population densities (less than five people per kilometer^2, sometimes less than one person per kilometer^2). They are important for a wide variety of reasons. They are vast storehouses of biodiversity; they protect critical watersheds and serve as controls against which we can measure the management of the more devastated hotspots; they are often the last places where indigenous people stand a chance of maintaining some semblance of their traditional lifestyles; they are increasingly important as recreational sites; and their spiritual and aesthetic value on an ever more overcrowded planet is sure to increase as well. These high-biodiversity major tropical wilderness areas are now few and far between, with the principal ones being found in the southern Guianas, southern Venezuela, and adjacent parts of extreme northern Brazilian Amazonia; parts of upper Amazonian Brazil, Colombia, Ecuador, Peru,

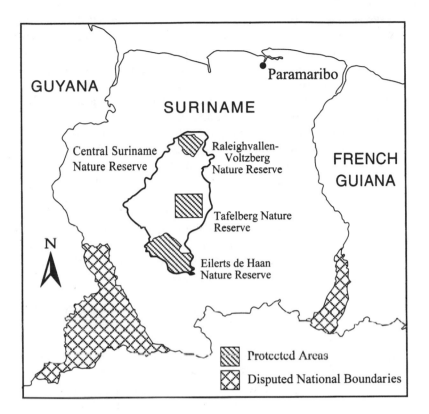

Figure 1.2   Central Suriname Nature Reserve.

and Bolivia; parts of the Congo Basin; and much of the island of New Guinea (Mittermeier et al. 1998).

An excellent example of conservation of these major tropical wilderness areas is the recent creation of a four million-acre reserve in the South American nation of Suriname. This new Central Suriname Nature Reserve, created by the government of Suriname in partnership with CI, alone encompasses more than 10 percent of the country and will serve as a cornerstone of conservation in the Guiana Shield wilderness area (see figure 1.2).

Megadiversity Countries

The final component of CI's global priority-setting strategy for biodiversity conservation is the concept of *megadiversity countries*, a concept first proposed by the senior author at the Smithsonian Institution's 1988 Biodiversity Conference (Mittermeier 1988). Initially, this concept grew out of an overview of the distribution of nonhuman primates, but later it was expanded to look at biodiversity in general

(Mittermeier and Werner 1990; McNeely et al. 1990; Mittermeier and Mittermeier 1992). This approach considers priorities for biodiversity conservation according to political units (i.e., sovereign nations), rather than by ecosystems, and is mainly intended to stimulate those countries that are rich in biodiversity to recognize their natural heritage as both a competitive economic advantage and a fundamental component of their culture and national character.

A country must meet several criteria to make the megadiversity list, but again the primary criterion is plant endemism, with data on mammals, birds, reptiles, and amphibians taken into consideration as well. In addition, emphasis is placed on ecosystem (beta) diversity, and all countries on the list must have marine ecosystems as well; no landlocked nations are accorded megadiversity status. This is in recognition of the tremendous diversity in the oceans, especially in terms of animal life. Furthermore, all countries but one (South Africa) have at least some tropical rain forest, in recognition of the fact that this is the richest of the terrestrial biomes. An exception is made for South Africa because of its tremendously high plant diversity and endemism and the fact that it has an entire plant kingdom (the Cape Floristic Kingdom) within its borders.

In its earliest versions, the megadiversity country analysis highlighted seven countries (Mittermeier and Werner 1990). That number subsequently increased to twelve (Mittermeier and Goettsch de Mittermeier 1992), and finally to seventeen out of a global total of 230 (see figure 1.3; Mittermeier et al. 1997). Remarkably, these seventeen countries—Brazil, Indonesia, Colombia, Peru, Mexico, Australia, Madagascar, China, the Philippines, India, Papua New Guinea, Ecuador, the United States of America, Venezuela, Malaysia, South Africa, and the Democratic Republic of Congo—account for more than a third of the world's nonfish vertebrates and two-thirds of the higher plants *as endemics*, and at least two-thirds and probably three-fourths or more of the planet's total terrestrial freshwater and marine biodiversity (see figure 1.3).

Priority-setting exercises by several other institutions further highlight the importance of the hotspots, the major tropical wilderness areas, and the megadiversity countries. For example, BirdLife International's recent analysis of Endemic Bird Areas (EBAs) identified Indonesia, Brazil, Peru, Colombia, and Mexico as having the highest number of EBAs (essentially avian hotspots), representing critical habitat for more than twenty-six hundred threatened species (Stattersfield et al. 1998). Furthermore, there is a high correlation between BirdLife's EBAs and the CI hotspots, with 141 of the 218 EBAs, or approximately 65 percent, occurring in hotspots we have identified. There is

Figure 1.3 Megadiversity Countries.

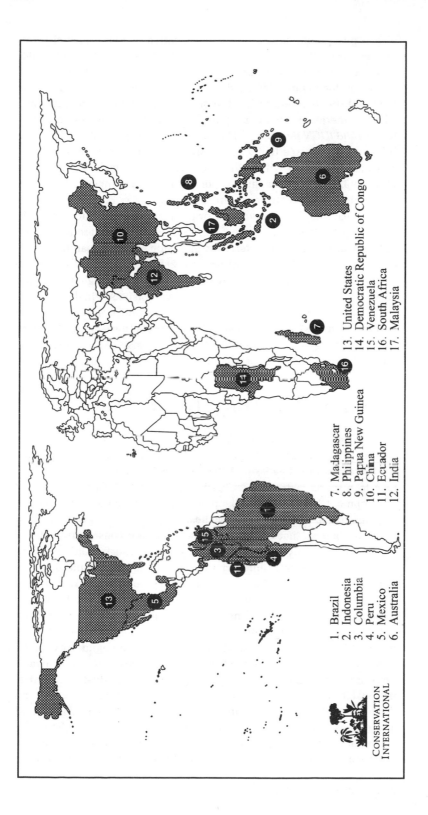

1. Brazil
2. Indonesia
3. Columbia
4. Peru
5. Mexico
6. Australia

7. Madagascar
8. Philippines
9. Papua New Guinea
10. China
11. Ecuador
12. India

13. United States
14. Democratic Republic of Congo
15. Venezuela
16. South Africa
17. Malaysia

CONSERVATION
INTERNATIONAL

also a strong correlation among the Centres of Plant Diversity that emerged from the recent study by the WWF and IUCN (1994, 1995, 1997), and both hotspots and megadiversity countries. Of the 125 critical areas for plants identified in the WWF and IUCN study, more than half the total fall within the borders of the megadiversity countries. Furthermore, considering biodiversity at greatest risk according to the *1996 IUCN Red List of Threatened Animals* (Baillie and Groombridge 1996), we see that 75–90 percent of the world's most endangered animals occur within the megadiversity countries (Mittermeier et al. 1997), and a comparable percentage of endangered terrestrial mammals and birds are found in the hotspots. Looking at WWF's recent Global 200 list, many more priority areas are included than in our hotspots, largely because of the criterion of representativeness and lack of quantitative criteria for inclusion on this broad list. However, if one looks at the highest priorities among the Global 200, those referred to as Extinction-Prone Areas, one finds a very high (though not surprising) correlation with the twenty-five hotspots. Finally, there is much overlap between the World Resources Institute's recent analysis of "Forest Frontiers" and the major tropical wilderness areas, the main difference being that they include the boreal forests of Russia and Canada in their analysis whereas CI does not (Bryant et al. 1997).

Recognizing the Critical Role of Protected Areas

Having identified these global priority areas for biodiversity conservation, the challenge is to pursue the best conservation strategy. This book examines the key issue: What approach should conservationists take to resource extraction? Clearly, we need to influence resource development where it does take place—and the chapters in this book highlight some promising approaches.

Yet we believe that one of the most important conservation strategies is, quite simply, protection itself, and the transformation of protection from an esoteric enterprise to an economically sound business opportunity. This leads us directly into the second of our key priority issues: recognizing *the critical role of protected areas* in global efforts to conserve biodiversity. Although the conservation movement originally began with a focus on endangered species and protected areas, much of the emphasis in recent years has been on community development, conservation-based enterprises, education and public awareness, and policy matters, such as best practices for industry and international treaties like the Convention on Biodiversity. There is no doubt that all of these are important and fit well in landscape-based approaches like the UNESCO Biosphere Reserve concept (first conceived of by conservation pioneer Michel Batisse of UNESCO), which has been an underpinning of CI's conservation strategy since the organization was created. However, if one looks carefully at the Biosphere Reserve concept, the key element is that of the nucleus—that core area of completely protected nature around which everything else revolves and upon which the overall integrity of the system ultimately depends. It is this truly critical element of biodiversity conservation that we fear the international community has downplayed too much in the last couple of decades, as people have sought to promote sustainable development outside of protected areas.

Why are protected areas so important, and why are they once again coming back into focus? First, although we need to maximize the diversity of life within the matrix of human-modified landscapes outside protected areas, the fact is that much of what remains of the world's biodiversity will be found in pristine (or as near to pristine as we can get) parks and reserves. The rarer, more specialized, slower breeding creatures with longer life spans, the creatures at the top of food chains with large home range requirements, and, in most cases, the largest, most charismatic megavertebrates that so fascinate us (e.g., elephants, rhinos, great apes, and big cats) and that represent the most significant elements in the rapidly growing global ecotourism industry, are likely to survive only in the best protected and largest parks and reserves. Furthermore, the simple fact is that, in areas of high human population density like much of Asia, Europe, and the industrialized parts of South America, and especially in the top-priority biodiversity hotspots mentioned earlier, all that is likely to remain, even relatively intact, is what has been set aside as parks or reserves. In many such areas, this is not something that will happen in the future; it already is the case, that is, all that remains of a number of ecosystem types and all that survives of certain key endangered species are in protected areas. Human habitations and cultivated land frequently go up to the very border of the unit in question, and the much discussed "buffer zone" many times represents nothing more than a theoretical construct in the design of a management plan.

In promoting protected areas to conserve global biodiversity, six important challenges must be met:

- Ensure that the integrity of all existing protected areas is maintained, including the buffer zones of those that still have them;
- Reinstate whatever buffer zones might once have existed, if possible, and allow for natural regeneration;
- Set aside new protected areas to provide as broad a representation of each nation's biodiversity as possible;
- Take a regional approach that seeks to connect existing and proposed protected areas through series of corridors;
- Allocate far more significant resources to protected area establishment and maintenance; and

- Invest far more than ever before in promoting ecotourism, scientific research, and other income-generating activities in protected areas so that they can realize their potential as income generators for the countries in question.

REFERENCES

Baillie, J., and B. Groombridge, eds. 1996. *1996 IUCN Red List of Threatened Animals*. IUCN Species Survival Commission, World Conservation Monitoring Centre and BirdLife International, Gland, Switzerland, and Cambridge, England.

Balmford, A., and A. Long. 1994. Avian endemism and tropical deforestation. *Nature* 372:623–624.

Bawa, K. S., and R. Seidler. 1998. Natural forest management and conservation of biodiversity in tropical forests. *Conservation Biology* 12(1):46–55.

Bibby, C. J., N. J. Collar, M. J. Crosby, M. F. Heath, C. Imboden, T. H. Johnson, A. J. Long, A.J. Sattersfield, and S. J. Theirgood. 1992. *Putting Biodiversity on the Map: Priority Areas for Global Conservation*. International Council for Bird Conservation, Cambridge, England.

Bowles, I., R. E. Rice, R. A. Mittermeier, and G. A. B. da Fonseca. 1998. Logging and tropical forest. *Conservation Science*. 280: 1899–1900.

Bryant, D., D. Nielsen, and L. Tangley. 1997. *The Last Frontier Forests: Ecosystems and Economies on the Edge*. World Resources Institute, Washington, D.C.

Conservation International. 1990a. *Biodiversity at Risk: A Preview of Conservation International's Atlas for the 1990s*. Conservation International, Washington, D.C.

———. 1990b. *The Rainforest Imperative: A Ten-Year Strategy to Save Earth's Most Threatened Ecosystems*. Conservation International, Washington, D.C.

Conservation International, Fundação Biodiversitas, and Socidedade Nordestina de Ecologia. 1994. *Prioridades para Conservação da Biodiversidade da Mata Atlântica do Nordeste*. Conservation International, Washington, D.C.

Conservation International, U.S. Man and Biosphere Program, ECO-SUR, USAID, Pasco Pantera Consortium, University of Florida, and Amigos de Sian Ka'an. 1996. *Evaluation of the Selva Maya*. Conservation International, Washington, D.C.

Costanza, R., R. d'Arge, R. de Groot, S. Farber, M. Grasso, B. Hannon, K. Limburg, S. Naeem, R. O'Neill, J. Paruelo, R. G. Raskin, P. Sutton, and M. van den Belt. 1997. The value of the world's ecosystem services and natural capital. *Nature* 387:253–260.

Erwin, T. L. 1983. Beetles and other insects of tropical forest canopies at Manaus, Brazil, sampled by insecticidal fogging. In *Tropical Rain Forest: Ecology and Management*, edited by S. L. Sutton, T. C. Whitmore, and A. C. Chadwick. Blackwell, Edinburgh. Pp. 59–75.

Foster, R. B., T. A. Parker III, A. H. Gentry, L. H. Emmons, A. Chicchón, T. Schulenberg, L. Rodríguez, G. Lamas, H. Ortega, J. Icochea, W. Wust, M. Romo, J. A. Castillo, O. Phillips, C. Reynel, A. Kratter, P. K. Donahue, and L. J. Barkley. 1994. *The Tambopata-*

Candamo Reserved Zone of Southeastern Perú: A Biological Assessment. RAP Working Paper no. 6. Conservation International, Washington, D.C.

Ganzhorn, J. U., B. Rakotosamimanana, L. Hannah, J. Hough, L. Iyer, S. Olivieri, S.Rajaobelina, C. Rodstrom, and G. Tilkin. 1997. Priorities for biodiversity conservation in Madagascar. *Primate Report* 48(1):1–81.

IUCN/UNEP/WWF. 1980. *World Conservation Strategy: Living Resource Conservation for Sustainable Development.* International Union for Conservation of Nature and Natural Resources, United Nations Environment Programme, and World Wildlife Fund, Gland, Switzerland.

———. 1991. *Caring for the Earth: A Strategy for Sustainable Living.* IUCN–The World Conservation Union, United Nations Environment Programme, and WWF–World Wide Fund for Nature, Gland, Switzerland.

Johnson, N., and B. Cabarle. 1993. *Surviving the Cut: Natural Forest Management in the Humid Tropics.* World Resources Institute, Washington, D.C.

McNeely, J. A., K. R. Miller, W. V. Reid, R. A. Mittermeier, and T. B. Werner. 1990. *Conserving the World's Biological Diversity.* WRI, World Conservation Union, World Bank, WWF-U.S., and Conservation International, Washington, D.C., and Gland, Switzerland.

Mittermeier, R. A. 1988. Primate diversity and the tropical forest: Case studies from Brazil and Madagascar, and the importance of the megadiversity countries. In *Biodiversity*, edited by E. O. Wilson. National Academy Press, Washington, D.C.

Mittermeier, R. A., and I. A. Bowles. 1993. The GEF and biodiversity conservation: Lessons to date and recommendations for the future. Policy Paper no. 1. Conservation International, Washington, D.C.

Mittermeier, R. A., and C. Goettsch Mittermeier. 1992. La importancia de la diversidad biológica en México. In *Mexico Confronts the Challenges of Biodiversity*, edited by J. Saruhkan and R. Dirzo. Comisión Nacional para el Conocimiento y Uso de la Biodiversidad, México, México.

Mittermeier, R. A., and T. B. Werner. 1990. Wealth of plants and animals unites "megadiversity" countries. *Tropicus* 4(1):4–5.

Mittermeier, R. A., C. Goettsch Mittermeier, and P. Robles Gil. 1997. *Megadiversity: Earth's Biologically Wealthiest Nations.* Cemex, Mexico City.

Mittermeier, R. A., N. Myers, J. B. Thomsen, G. A. B. da Fonseca, and S. Olivieri. 1998. Biodiversity hotspots and major tropical wilderness areas: Approaches to setting conservation priorities. *Conservation Biology* 12(3):1–5.

Myers, N. 1983. *A Wealth of Wild Species: Storehouse for Human Welfare.* Westview Press, Boulder, Colorado.

———. 1988. Threatened biotas: "Hot-spots" in tropical forests. *Environmentalist* 8(3):187–208.

———. 1990. The biodiversity challenge: Expanded hot-spots analysis. *Environmentalist* 10(4):243–256.

Parker, T. A., III, et al. 1993a. *The Lowland Dry Forests of Santa Cruz, Bolivia: A Global Conservation Priority.* RAP Working Paper no. 4. Conservation International, Washington, D.C.

Parker, T. A., III, et al. 1993b. *A Biological Assessment of the Columbia River Forest Reserve, Toledo District, Belize.* RAP Working Paper no. 3. Conservation International, Washington, D.C.

Parker, S. P., ed. 1982. *Synopsis and Classification of Living Organisms.* McGraw-Hill, New York.

Parker, T. A., III, and B. Bailey, eds. 1991. *A Biological Assessment of the Alto Madidi Region.* RAP Working Paper no. 1. Conservation International, Washington, D.C.

Parker, T. A., III, and J. Carr, eds. 1992. *Status of Forest Remnants in the Cordillera de la Costa and Adjacent Areas of Southwestern Ecuador.* RAP Working Paper no. 2. Conservation International, Washington, D.C.

Pimm, S. L., J. L. Gittleman, J. Russell, and T. M. Brooks. 1995. The future of diversity. *Science* 269:347.

Schulenberg, T. S., and K. Awbrey, eds. 1997a. *The Cordillera del Cóndor Region of Ecuador and Peru: A Biological Assessment.* RAP Working Paper no. 7. Conservation International, Washington, D.C.

————. 1997b. *A Rapid Assessment of the Humid Forests of South Central Chuquisaca, Bolivia.* RAP Working Paper no. 8. Conservation International, Washington, D.C.

Sizer, N. 1996. *Profit without Plunder: Reaping Revenue from Guyana's Tropical Forests without Destroying Them.* World Resources Institute, Washington, D.C.

Sizer, N., and R. Rice. 1995. *Backs to the Wall in Suriname: Forest Policy in a Country in Crisis.* World Resources Institute, Washington, D.C.

Stattersfield, A. J., M. J. Crosby, A. J. Long, and D. C. Wege. 1998. *Endemic Bird Areas of the World: Priorities for Biodiversity Conservation.* BirdLife International, Cambridge, England.

Walter, K. S., and Gillett, H. J. 1997. *1997 IUCN Red List of Threatened Plants.* IUCN–World Conservation Union and World Conservation Monitoring Centre, Gland, Switzerland, and Cambridge, England.

Wilson, E. O. 1985. The biological diversity crisis: A challenge to science. *Issues in Science and Technology* 2:20–29.

Wilson, E. O. 1992. *The Diversity of Life.* Harvard University Press, Cambridge, Mass.

Wilson, E. O., ed. 1988. *Biodiversity.* National Academy Press, Washington, D.C.

World Conservation Monitoring Centre. 1992. *Global Biodiversity: Status of the Earth's Living Resources.* World Conservation Monitoring Centre, Cambridge, England.

WRI, IUCN, and UNEP. 1992. *Global Biodiversity Strategy: Guidelines for Actions to Save, Study, and Use the Earth's Biotic Wealth Sustainably and Equitably.* World Resources Institute, Washington, D.C.

WWF and IUCN. 1994. *Centres of Plant Diversity: A Guide and Strategy for their Conservation. Volume 1. Europe, Asia, Africa, South West Asia and the Middle East.* IUCN Publications, Cambridge, England.

————. 1995. *Centres of Plant Diversity: A Guide and Strategy for their Conservation. Volume 2. Asia, Australasia and the Pacific.* IUCN Publications, Cambridge, England.

————. 1997. *Centres of Plant Diversity: A Guide and Strategy for their Conservation. Volume 3. The Americas.* IUCN Publications, Cambridge, England.

2 Private-Sector Participation in
Infrastructure Development

Everett J. Santos

For most of this century, growth in the developing world has been hamstrung by restrictions on the development, financing, and opera tion of its infrastructure by the state. Within the last decade, a new paradigm has been adopted by more and more of the developing world: the privatization of infrastructure. The impact on development, economic growth, and the environment will be vast. To fully understand how the world will change as a consequence of this para digm shift, it is necessary to grasp why the transition to private provisioning of infrastructure has become necessary and how it will be implemented.

Need for Infrastructure in the Developing World

Infrastructure is the substructure of development. It is the determining factor of a country's economic and social development. Similarly, it advances and prescribes a nation's industrial and commercial competitiveness. Access to electricity is necessary both for industries to compete and for communities to participate in the benefits of the modern world. Telecommunications has become the nerve network enabling efficiencies in production. The quality and expanse of a country's transportation will control its ability to participate in modern industrial competitive markets. The efficient movement of people and goods is fundamental to social and economic integration. A society's health is to a great extent determined by a community's access to clean water, waste disposal, and sewage facilities.

Until recently, much of the world presumed that providing infrastructure was the exclusive responsibility of the state, or at least that the private sector could not be expected to play a significant role in such provisioning. With few exceptions, governments were expected to assess infrastructure needs, evaluate alternatives, price services, regulate provisioning, finance projects, generate savings, and maintain assets. Add to that litany of responsibilities the task of balancing the interest of the consumer against the costs of providing infrastructure services economically and in an environmentally sustainable manner, and the enormity and absurdity of the challenge accepted by governments can be understood. It is no wonder that infrastructure provisioning was so shortchanged in the bargain.

It is against this backdrop of inadequate services and a flawed structure that the world began to search for a better approach to provisioning infrastructure. While still recognizing that government has an important role to play, country after country has declared a need and an intention to have the private sector undertake a larger role in infrastructure development.

The factors leading to the paradigm shift are many. The collapse of the command economies of Eastern Europe are an obvious factor but probably not singular in their importance. Too many other factors were also covertly building pressure on the status quo in developing countries. The debt crisis, the operating inefficiency experienced by almost all publicly owned enterprises, the increased awareness and demand of customers dissatisfied with the state of infrastructure services, financially strained treasuries that could no longer sustain the demands for increased investments, even technological breakthroughs all contributed to the push to privatize.

The need, even the economic demand, for infrastructure services in the developing world is colossal. Economists would claim that investments—assuming a benign economic environment—should be commensurate with the economic demand for the services to be provided by the investment. According to the World Bank, no less than $2 trillion will be needed and possibly invested during the next ten years to fund power, telecommunications, and transport projects in less-developed countries (LDCs). Many LDCs are already experiencing remarkable results from moving infrastructure to the private sector: New foreign investments, the deepening of capital markets, and an improved standing in the international banking communities are all the results, in part, of involving the marketplace in the provisioning of infrastructure.

LDCs are, in development terms, far below where the United States stood about fifty years ago, at the end of World War II, and many suffer conditions that are below what the United States experienced a century ago. The rate of population growth in LDCs has been much higher than the rate at which the developing world has been able to

Table 2.1 Infrastructure Conditions in Developing
Regions

	Percent of Population with Service	
Region	Water and Sanitation	Electricity
Africa	42%	22%
Asia	52%	60%
Eastern Europe	82%	97%
Middle East	68%	43%
Latin America	66%	55%

maintain coverage of electricity, water, and sanitation for its population. The number without these services is larger today than during
the 1940s, even though the absolute number and proportion of people
with access to these services has increased.

The Failure of State-Run Enterprises

Since the end of World War II, the public sector in most countries
has owned, controlled, and operated the so-called strategic industries.
Immediately after the war, the perceived inability of the private sector
to finance the infrastructure needs of the LDCs and ideological prefer-
ence to have the state control the "commanding heights" led those
LDCs that did not have state control of infrastructure services to na-
tionalize them. The move extended far beyond the classic areas of
infrastructure—power, transportation, water and sanitation, and telecommunications—to practically all significant sectors of these countries' economies. By the early 1970s, Mexico had 1,000 state-owned
enterprises; Chile, 650; Venezuela, 400; Brazil, 200; Indonesia, 212,
the Philippines, 106; and Thailand, 63. Government mismanagement
has left the LDCs with distorted patterns of demand, confused access
to services, inefficient operations, and severely deteriorated capital
assets, which have suffered from years of mistreatment and poor
maintenance.

The crisis faced by LDCs in the 1980s led to a re-examination of
public-sector operations. Utilities had incurred large sums of debt
that governments were obliged to repay. Unfortunately, most public-
sector infrastructure enterprises had insufficient earnings to meet
these debt obligations. The servicing of obligations in some cases
reached critical levels and contributed to the debt crisis of many Latin
American and African countries. The combined debt of state enterprises in Latin America's largest economies in 1983 was $37.6 billion

in Argentina, $53.3 billion in Brazil, and $50.6 billion in Mexico. By the early 1980s, state enterprises represented half of Latin America's public-sector deficit, compared with 25 percent ten years earlier.

Although there are many reasons for operating inefficiency in state-owned enterprises (SOEs), one is overriding: lack of accountability. Major regulatory changes that encourage competition and market-driven decisions should be fostered. Domestic and foreign capital markets should be designed to encourage long-term investment planning. Infrastructure service itself should be made to operate in a competitive market. And finally, government must change its role from that of provider to that of regulator.

Central government authorities have often subsidized infrastructure services under the pretext of assisting the society's disadvantaged. It was believed that subsidies assist the poor and promote growth and economic development. But much of what is passed off as largesse to the poor in fact is nothing more than support for middle-class consumption. For example, few of the truly poor in many of the countries that try to justify electricity subsidies actually receive electricity, that is, are wired. Subsidies thus benefit the wrong people and promote waste and inefficient use of valuable resources, with the attendant negative impact to the environment.

Access to infrastructure is heavily biased toward those living in the cities, which often do not house the most disadvantaged populations. For example, while in most of the developing world's urban areas 88 percent of the population has access to safe drinking water, only 39 percent of those in rural areas do. In around half the countries for which data is available, 30 percent or more of urban households are not connected to electricity, while three out of every four rural homes are not connected. About 15 percent of the water and 30 percent of the electricity produced is wasted or stolen before it reaches customers. In about half of the LDCs, there is an average of one telephone line to every one thousand people. Maintenance is overdue on between 60 and 70 percent of the developing countries' roads.

Governmental Financing

Most private-sector infrastructure financing in LDCs was provided by specialized domestic governmental agencies, export credit agencies (ECAs), commercial banks, or multilateral development banks (MDBs)—such as the World Bank, International Finance Corporation (IFC), and regional development banks (RDBs)—until the end of the 1980s. In providing monetary assistance to state-run operations and to governments intent on retaining a dominant role in the sector, the financiers have become a party to the inefficiencies of the sector. For a variety of reasons, they will no longer be able to provide the level of funds needed for infrastructure development:

- Government agencies have large fiscal deficits, and their more pressing social spending needs have absorbed the little that had been available for infrastructure.
- Multilateral organizations cannot possibly meet the enormous requirements ($2 trillion) of the sector. Also, increasing criticisms of the multilateral financing community and the fiscal deficit difficulties faced by some of the MDBs' major contributors will force the MDBs to reduce the proportion of assistance they can provide the sector.
- Export credit agencies have historically demanded governments' full faith and credit backing for their loans. As the sector is privatized, ECAs will be forced out or will need to change their operating procedures. Theoretically, ECAs should be willing to finance private infrastructure projects, but this requires a willingness to accept project risk. Until recently, no such inclination existed.
- Among the traditional sources of infrastructure financing, the commercial banks have the most flexibility to increase their relative position. Although countries cannot undergo bankruptcy procedures, under unfavorable macroeconomic conditions they can "act bankrupt." To protect themselves, banks have turned to project financing backed by reliable revenue streams instead of country guarantees. This should permit commercial banks to intensify their role as financiers of infrastructure projects over the coming years.

The multilateral lending institutions have been criticized by a wide array of parties. Environmentalists are concerned about the impact of MDBs' infrastructure development programs and policies on the environment. The political Left has questioned the effectiveness of addressing the needs of the poorer segments of the LDCs of those programs and policies. The political Right has opposed the overemphasis on government-directed solutions, which hamper the private sector in the developing world. The list of critics goes on. The same criticisms can be levied with the same degree of relevance at government participation in infrastructure development itself. Whereas everyone agrees that there are enormous infrastructure deficiencies throughout the developing world, how governments can best address those needs is a basic question.

The criticisms are valid, but there is a more fundamental question: What should be the role of MDBs and governments in a reoriented world where private capital and markets determine what will get done and how? What should be the objective of MDBs and governments? Does the existing structure achieve these objectives? While governments are being "reinvented" and corporations are being "re-engineered," a legitimate question exists as to whether government policies designed for the chaos of the postwar 1940s are appropriate for a world with distinctively different economic circumstances. The world has become a global marketplace. The international capital markets are free and open and expanding. And in a rapidly changing world, there appears to be a real question as to the appropriate role of governments and their institutions. As an editorial in the *Economist* stated,

> The assumption on which the World Bank was founded in 1945—that the best way to help impoverished countries overcome chronic problems was to inject enlightened multilateral capital into them, usually through their governments—has come to look less convincing. International private capital markets have expanded enormously, on the one hand, and private sector entrepreneurs have begun to elbow aside state controls (*Economist* 19) on the other.
>
> The reality is that governments are responding to a new world by adapting their strategies to rapid changes and increasingly resorting to private-sector entrepreneurship, and by facilitating a more efficient use of the international capital markets.

Infrastructure Financing through Capital Markets and Foreign Investors

For the private sector to succeed in infrastructure development, it must have access to sufficient financial resources. Only then can it effectively own, operate, and sustain infrastructure services, update industries, and introduce new technologies. This financial support should come from both foreign investors and efficient domestic capital markets. Otherwise, a real danger exists that the trend to privatize infrastructure in the LDCs will be stalled, sidetracked, or even reversed.

Increased investment opportunities through the privatization of public-sector industries—previously closed to foreign (and domestic) investors—has partially stimulated the recent trend toward the internationalization of the capital markets. Privatization has helped reduce government deficits and inflation, thus stabilizing economies and creating an environment conducive to growth of private-sector investment. But the privatization of very large companies in capital-scarce countries cannot thrive without foreign resources.

Shifting infrastructure projects into the private sector may be an important route for developing capital markets, as there are important synergies between these forces. The absence of infrastructure projects

capable of issuing stable, long-term private securities has undoubtedly repressed the development of such securities. In turn, the absence of stable, long-term securities hampers the broadening and
deepening of securities markets. It is ironic that some governments
justify their intervention in infrastructure services because of the absence in their countries of a mature capital market. Yet the very act
of diminishing the growth of private infrastructure deprives the markets of the means to expand. Increasing the amount, variety, and size
of financial instruments available to investors and the number and
diversity of financial institutions needed to provide them serves to
augment the options available to these investors. These changes contribute to enhancing the breadth and depth of capital markets.

LDC Difficulties in Raising Funds

Infrastructure ventures formed from the privatization of formerly public enterprises need to recover from the effects of government mismanagement, frequently including poor financial performance when
they were operated by the public sector. Reversing a history of mismanagement is a formidable task. A private venture may have accounts receivable problems simply because its only client, a public-
sector company, had financial problems. In many cases, privatized
companies and new ventures formed from governments opening sectors to investors have no track record from which their own performance may be judged, at least none that would be sufficient to attract
investors.

Regulatory Framework

Although private competitive markets can go a long way in helping
finance infrastructure projects, no country can afford to completely
stand back and assume that the private sector will perform every role
necessary to assure the nurturing of sustainable development. Experience has shown that markets can go awry and that, without an appropriate regulatory environment, abuses will occur in the operation and
pricing of business concerns, the sale of securities, and the raising of
capital. Indeed, only the most enlightened business concerns will
adopt environmentally friendly practices absent any strong financial
or political reasons to do so. There are, however, universally applicable precepts on how to best regulate markets and foster competition,
promote social justice, and provide environmentally sustainable development.

The same is true of environmental regulations. To assure adherence
to sound environmental policies, business must be given financial

"incentives" in the form of either making it costly to pollute or profitable to enhance the environment. Governments were never very good at developing projects in an environmentally sustainable manner. A look at Eastern Europe's air quality or the emissions at most government-owned facilities in the developing would prove that case without any doubt. In fact, private industry, in its effort not to waste resources, may be more environmentally benign.

Every age has its critical developments that dictate how the next generation will live and that pose their own set of additional issues for resolution. The end of World War I was such a time, and the failure to set in place a fair and lasting peace led to the catastrophe of World War II. Similarly, the end of that war set in motion the development mode of command economies and an overreliance on the state to address the infrastructure needs of developing and restructuring countries. Unfortunately, that model fell short of meeting the challenge of how to provide universal coverage of critical infrastructure needs.

The world is now accepting a new paradigm for infrastructure development. If the new paradigm is up to the challenge, within the next two to three decades universal coverage will be provided to everyone on this planet. What this will mean for the poorest in our world is access to the living standards that the advanced economies have taken for granted, as a matter of right, for the last fifty years. But it also will impose a formidable and unprecedented challenge in terms of assuring that universal access does not compromise the environment. The challenge requires that governments and the private sector assume different roles. The environment will need to be safeguarded by more rigorous standards than ever before. Governments will have to enforce those standards. More important—as the mantle of infrastructure development is passed to the private sector—economic parameters within which the private sector will play out its role must be designed to guarantee that the incentives they create will promote environmentally sustainable development.

Both a country's environment and infrastructure will determine the quality of life that any society can provide its citizens. Unfortunately, the environment and infrastructure development have been characterized as foes, where the interest of one compromises the interest of the other. Yet nothing could be further from the truth. Sustainable development requires that a country develop its infrastructure so that its very environment can be protected from the assault that invariably comes when a people suffer from poverty and are forced to sacrifice the quality of their environment just to survive. It is not coincidental that the countries with highest incomes are also the countries with the most advanced infrastructure and levels of environmental awareness.

3 Partner or Pariah

Public Perceptions and Responses to the Extractive Industries

Lisa Jordan
Christopher H. Chamberlain

On April 14, 1998, millions of readers of the *New York Times* turned to an unusual advertisement among the typical pitches for lingerie and airfare deals. An "Investor's Advisory Concerning a Controversial Oil Project in the Colombian Amazon," the ad read: "Why Occidental's oil project is a death sentence for the U'wa." The full-page ad explained that Colombia's indigenous U'wa people were threatening to commit mass suicide, repeating a legendary U'wa action in the face of seventeenth-century colonialism, if Occidental Petroleum Company proceeded with a major drilling project on U'wa lands. "Consider what will actually happen if Occidental and its partners push forward against the will of the U'wa," the ad warned. "The U'wa have a simple request: the right to say no." Beneath, a photo of a forlorn U'wa girl in the ad makes its message clear: Potential investors beware.

This poignant appeal for the life and land of an indigenous culture undoubtedly shocked many of the *Times*'s readers. But the struggle of the U'wa is hardly uncommon. In recent years, the extractive industries—mining, oil and gas, and timber harvesting—have come under increasing attack as their impacts on the environment and local communities are more widely understood. Affected peoples, environmental pressure groups, labor movements, consumers, and socially conscious investors have all devised sophisticated strategies—and often formed intensive partnerships, as in the case of the U'wa—to reform or halt the destructive impacts of the extractive industries.

Besieged by political pressure, financial hardship, and legal exposure, the extractive industries have been deeply affected by the public

37

campaigns waged against them. Fearing the consequences of becoming pariahs in the public eye, the industries have developed sophisticated approaches in order to address an increasingly mobilized, hostile public.

The root of these complex and contentious politics lies in the serious impacts of the extractive industries. Threats to the global climate, biodiversity, and the environment occur in conjunction with devastating impacts on the social, cultural, and economic sustainability of project-affected communities. While these costs are great, the massive profits that can be reaped from the extraction of minerals, oil, gas, and timber only intensify the battles and complicate the alliances among industry, government, affected peoples, and environmentalists.

This chapter provides an overview of the political battles that stem from environmentally and socially destructive resource extraction. The authors illustrate some of the ways in which the public—including local affected communities, nongovernmental organizations (NGOs), and consumers in the developed countries—express their concerns and seek to reduce the social and environmental costs of the extractive industries. We describe the "pressure points" that the public uses to stop or remedy destructive extraction projects, such as targeting the financial institutions that provide a project's financing and underwriting. In particular, we stress the impact on public lenders (including bilateral government agencies and international financial institutions like the World Bank Group) of public campaigns over the last two decades. We also evaluate the effectiveness of the industries' responses to public pressure, and we assess the future political landscape within which the extractive sector might operate.

The Public Concern

Project Underground, a U.S.-based NGO, produces *Drillbits and Tailings*, a bimonthly newsletter that alerts readers to problematic extractive operations around the world. Each issue contains a section, "Hotspots," which features news of extractive projects that have become the target of public action and concern. The extensive number of "Hotspots" that have appeared in *Drillbits and Tailings* over the last two years illustrates the public's growing concern with the extractive industries.

The April 1998 "Hotspots" included news from Alaska, China, Cuba, Lesotho, Madagascar, Montana, Nova Scotia, Peru, Philippines, Saskatchewan, Sierra Leone, Utah, Venezuela, and Wisconsin. The following items, excerpted from the April edition, indicate the scope of the controversy and contention that surround extractive projects throughout the world.

- **Philippines**: The use of mercury in Diwalwal, a gold mining area in the southern Philippines, has been banned after a check on residents showed at least thirty-six children and four government health workers suffered mercury poisoning, a government official said on March 21. Meanwhile, and in a contrary development, foreign mining companies have scaled down their operations in the Philippines because of uncertainties over a new tribal law and mounting environmental concerns, industry leaders said that same week. The Indigenous People's Rights Act (IPRA), a new Philippine law giving tribal groups rights over ancestral lands, has cast a shadow over Manila's liberalization policy, Horacio Ramos, Mines and Geosciences Bureau chief, told a news conference (*Associated Press*, March 21, 1998; *Reuters*, March 17, 1998).

- **Utah**: Federal prosecutors have filed a lawsuit against Texaco, the New York–based oil multinational, for polluting the San Juan River on indigenous Navajo lands in southeastern Utah. The suit cites eighty-eight different incidents of pollution. Additionally, prosecutors contend the company did not comply with environmental laws regarding cleanup plans and that it did not notify environmental regulators of some spills. Government lawyers say the spills alone could amount to almost $2.3 million in fines for Texaco (*Associated Press*, March 28, 1998).

- **Venezuela**: The Venezuelan College of Sociologists and Anthropologists has gone on the offensive against mining companies operating in the Imataca Rainforest Reserve in southeastern Bolivar State. Along with environmental and indigenous groups, the college introduced a Constitutional Injunction last May (1997) against Presidential Decree 1850, which permits gold mining activities in 40 percent of the 3.2 million hectares (7.9 million acres) of the Imataca Rainforest Reserve, which is considered fragile and is home to 10,000 Pemon Indians (*VHeadline/VENews*: March 23, 1998).

- **Wisconsin**: Republican Governor of Wisconsin Tommy Thompson has announced that he will sign the Sulfide Mining Moratorium Bill (SB3) this month. The bill will require companies to provide one example of a sulfide mine that has operated for ten years (and been closed for ten years) without polluting the surrounding environment, before any company can get a mine permit. It is perceived as a huge victory for the grassroots movement opposed to the proposed Crandon sulfide mine, to be operated by Rio Algom, because providing one example is an insurmountable hurdle (*Mid West Treaty Network*, March 25, 1998).

Project Underground, which formed in 1996 to monitor and conduct advocacy on the extractive industries, is not alone—similar organizations exist in Europe, Australia, and several of the developing countries. For example, the United Kingdom–based Minewatch focuses solely on monitoring the social and environmental impacts of the extractive industries. In Africa, environmental and development NGOs are establishing networks to target destructive extraction projects. In the Philippines, environmental groups are hiring staff to assist indigenous groups that are struggling to protect their rights in the face of extractive projects. In several cases, NGOs have even formed to target one specific company. The United Kingdom–based Rio Tinto, or RTZ, has the dubious honor of receiving the undivided attention of Partizans, an NGO that formed in response to the company's egregious environmental record. Royal Dutch Shell is similarly targeted by coalitions like the International Roundtable on Nigeria.

Virtually all these initiatives have occurred over the last five years, which suggests that the public's resentment towards the extractive industries has increased substantially. Mining, logging, and petroleum companies seem to be emerging in the public eye as the new "pariah" industries, similar to the nuclear power industry, which first earned this distinction in the 1970s, and the military industrial complex, which attracted public ire in the 1980s. The industries' emerging pariah status can be attributed in part to the dramatic documentation of their negative impacts on the environment and indigenous peoples; a growing conception in the Western world of the limitations of the earth's resources; the industries' perceived collusion with repressive military regimes around the world; and the growing public backlash against globalization.

This recent rise in public concern surrounding the extractive sector has also been fueled by the industries' rapid proliferation into the developing countries. The "frontier areas" of Latin America, Sub-Saharan Africa, and central and southeast Asia were once largely off limits to extraction—due to a combination of Cold War politics, financial risks, and technical constraints. Now, extractive companies are saturating previously untouched areas as they seek to replace the dwindling resources of their traditional locales. In 1996, the extractive industries spent $4.6 billion in exploration, with Latin America, Central and West Africa, and Southeast Asia as the main areas of expansion (Control Risks Group 1997). In Africa alone, $552 million was invested in nonferrous exploration in 1996, up from $199 million in 1994 (Da Rosa 1998).

The Public Response

As public concern over the extractive sector has increased, the public's reaction to the sector's negative impacts has intensified—and become more sophisticated. While project campaigns almost always begin with protests by the affected community, new methods of public pressure have led to complex alliances that involve actors in many different political arenas and that take advantage of modern technology and the mainstream media. The Occidental/U'wa case is typical in that it is conducted by an alliance of international advocacy organizations and the locally affected community and involves a strategy that applies simultaneous pressure on a variety of leverage points essential to Occidental's project.

The Public Response to Mitsubishi in the Sarawak

The campaign against Mitsubishi that has been spearheaded by the Rainforest Action Network (RAN) illustrates the methodology behind multilayered public campaigns. In the 1980s, an international campaign was formed to protest the rapid pace of the corporation's clear-cut logging in Sarawak, Malaysia. The campaign highlighted the plight of the indigenous Penan people, who were losing their land and livelihood as a result of the logging. Throughout the latter half of the decade, the Penan fought for recognition of their land rights, lobbying the Malaysian government and prominent families who controlled the Malaysian timber industry. Research conducted by many groups, including RAN, identified Mitsubishi as a majority shareholder in a company called Daiya Malaysia, which operated a ninety thousand-hectare logging concession in the Sarawak. Mitsubishi was one of the few corporations with an internationally recognized identity that was operating in Sarawak.

By 1988, the world was aware of both the plight of the Penan and the logging of the Sarawak rain forest—which was taking place twenty-four hours per day. Representatives from the Penan met with European and U.S. politicians, calls were made for a boycott of Malaysian logs, and appeals for international support and protection of the Penan became frequent. Subsequently, municipalities throughout Europe banned tropical forest imports, and consumer boycotts of unsustainably produced logs were launched. The European Parliament introduced a resolution that condemned the Sarawak situation and expressed concern about the future of the Penan.

In the face of global criticism, the Malaysian government conducted damage control, initiating stakeholder dialogues within the auspices of the United Nations and the International Tropical Timber Organization. Malaysian embassy officials met frequently with NGOs and other governments to counter the stories put forth by independent investigators and international journalists. But just as the Malaysian government's public relations intensified, the logging of the Sarawak increased, and many Penan representatives were imprisoned. The failure of the global Sarawak campaign to protect the Penan people prompted RAN to launch its Mitsubishi campaign.

RAN's effort to organize a boycott against Mitsubishi began in 1990 and was aimed at preventing the company from both conducting logging activities and purchasing timber products from other environmentally and socially destructive businesses. RAN publicized the details of Mitsubishi's vast and complex corporate structure, known as the Mitsubishi Group. This corporation, RAN reported, consisted of 160 companies that included gas, oil, mining, and nuclear operations.

RAN adopted techniques used by pressure groups like Greenpeace in order to attract attention to the Mitsubishi Group. It conducted street protests outside the corporation's headquarters and automobile dealerships and also targeted public events Mitsubishi sponsored, like the Grand Prix and the Super Bowl. RAN splayed giant banners over freeways and hung them atop high-rise office buildings. As in the U'wa case, RAN placed a full-page ad in the *New York Times* that featured the names and photographs of Mitsubishi executives, labeling them as among the world's greatest rain forest destroyers. RAN also produced a video, organized other environmental groups to support the campaign, coordinated door-to-door canvassing, distributed materials to primary and secondary schools, and hired scientists to review sustainable alternatives that could meet the consumer demand for forest products. RAN also began a corporate responsibility campaign, urging other corporations to condemn Mitsubishi's actions and urging individual investors to raise environmental concerns at Mitsubishi's shareholder meetings.

The Mitsubishi consumer campaign, which continues today, was possible to launch because of the company's familiarity to consumers worldwide. Yet the complexities of the corporation's structure have created significant difficulties in targeting the company as a whole. Thus, it has been able to limit its boycott exposure to a few of its arms. Mitsubishi Electric America and Mitsubishi Motor Sales, two of the hardest-hit subsidiaries of the corporation, have since signed an unprecedented agreement with RAN to review the environmental impacts of their activities. They have pledged to end use of old-growth forest products and to phase out the use of tree-based paper and packaging products by 2002. They have also created a "Forest Community Support Program" to restore and preserve the world's remaining ancient forests and to support their indigenous cultures.

Nevertheless, this agreement does not affect the actions of Mitsubishi's companies outside the United States. And in return for these concessions, RAN agreed to cease direct action against the two subsidiaries—a bargain that may severely limit RAN from carrying out effective future activities against the corporation as a whole.

While RAN has achieved significant results in its Mitsubishi campaign, this case also illustrates the difficulty of conducting campaigns against complex corporations that operate in the international arena. And indeed, many of the lesser-known actors that make up the extractive industries are even more resilient to public pressure than high-profile multinationals like Mitsubishi. While a well-recognized name can be a viable target for consumer action, it is highly unlikely that consumer-driven responses against a bauxite company or a no-name petroleum producer could be mobilized. In the Sarawak case, the public has been unable to directly target the other companies that are harvesting the rain forest alongside Mitsubishi, due to the companies' relative obscurity. So while the public often decries multinational corporations like Mitsubishi as the "problem" behind the negative impacts of the extractive sector, the smaller actors can often commit similar or worse abuses while enjoying a greater degree of impunity.

The Cutting Edge: Targeting Private Financial Mechanisms

Boycotts and media protests are direct and highly visible tactics. However, the public has also been developing a broader, more diverse set of strategies to address the growing legacy of the extractive industries. For example, Western NGOs have started to scrutinize the private financial services sector, which is integral to extractive project financing, as a potential entry point for public pressure. NGOs have begun to "map" the financial services industry, focusing on project finance mechanisms like mutual funds, insurance underwriting, and commercial banks as potential leverage points for advocacy (Buffet, Ganzi, and Seymour 1998). Using this strategy, future campaigns may increasingly target not only the sponsor of a destructive extraction project but also its commercial lenders, guarantors, and individual and institutional investors.

In *Leverage for the Environment: Strategic Mapping of the Private Financial Services Industry*, the World Resources Institute (WRI) defines four categories of pressuring, or "leveraging," that the public may be able to use in influencing development projects. The first, and most significant, is "bottom-line leverage," in which financial decisionmakers perceive that environmental concerns could affect their company's profit margin.

Indeed, there is a growing body of empirical evidence that suggests that this could be an effective strategy. Three researchers recently concluded that in some developing countries, individual firms' market values increased if they were explicitly recognized or rewarded for superior environmental performance. Conversely, citizen complaints against individual firms resulted in a decrease in their value (Dasgupta, Laplante, and Mimingi 1998).

guidelines, and ensuring citizen access to critical information and decisionmaking processes. NGOs have observed that IFIs like the IFC operate in this manner because they compete commercially with private financiers and must therefore adhere to marketplace norms, such as business confidentiality and speed. However, some NGOs claim that the IFIs use the argument of market norms as an excuse to limit their own exposure to public scrutiny.

Due to intensive public pressuring campaigns, as in the preceding OPIC example, all the IFIs have established policy frameworks to ensure that their activities "do no harm" to the environment or to local communities. NGOs, in turn, have learned to use the IFIs' policies and procedural mechanisms, like the requirement to conduct an environmental impact assessment (EIA), for projects with fewer negative impacts on the environment and local communities.

The Industry Reaction

Whether their motive is winning over the affected public, controlling the damage to their corporate identity, or preventing legal and political battles, the extractive industries react to public concerns and actions in a number of ways. The extent of a company's response is based on many factors, including the public's effectiveness in campaigning against a project, the gravity of the project's impacts, the degree to which the company has followed host country and international regulatory requirements, the host country's political and socioeconomic situation, and the company's relationship with the host government.

We will not discuss in detail the numerous tactics the industries use to respond to public criticism, but it is important to note that there is a continuum of corporate practice, which ranges from good-faith efforts to mitigate destructive impacts and engage affected communities, to hollow public relations stunts and unethical pressuring tactics. Well-known responses to public pressure, such as establishing voluntary codes of corporate conduct, as Shell has done, or initiating local community development programs, as Freeport Mc-MoRan and numerous others have done, receive mixed reviews from the public and thus are not very effective in stemming public criticism.

As a whole, the extractive industries have not been effective in addressing the growing public concern about their activities. One reason may be that virtually all responses have been *reactive*, in that companies have sought merely to solve problems that arise throughout an individual project's life cycle. Furthermore, companies tend to react *locally*, assuming that once the problems in a project's immediate area have been solved, the pressure from the global arena is likely to diminish. Few companies seem to have adjusted their overall corporate practices in order to avoid similar opposition and problems in future projects.

The industries have done little to proactively convince the public that they should not be the new corporate pariahs. Each high-profile battle, like Shell in Nigeria, or the Occidental/U'wa case, reinforces their negative corporate identity in the public eye. Meanwhile, the vast riches that the companies reap only serve to worsen their collective image as greedy entities who profit wildly at the expense of the earth's resources, indigenous peoples, and local communities.

In a speech to the Northwest Mining Association, a senior manager of the mining giant Placer Dome warned his colleagues of the growing public fear and resentment towards the sector. "The mining industry has the capacity to influence the world's judgment and sustain our welcome in the long term. The question is whether we have the collective inspiration and will to do so" (Control Risks Group 1997). What can the extractive industries do to allay the public's resentment and sustain their welcome? Clearly, they must proactively seek to address the root causes of public concern, which are repeated throughout the many hotspots around the world.

First, the industries should agree that some areas, despite their potential for mineral riches, must be off-limits to exploration and extraction. Regions rich in biodiversity, primary frontier forests, and lands that indigenous peoples depend on for cultural identity and economic livelihood should not be exploited. The industries could work with existing public entities to decide which areas would be best protected in order to preserve the health of the planet and protect its cultures. In seeking to achieve these goals—which would surely boost their international stature—the industries could agree to honor exclusionary guidelines that are also accepted by NGOs and the public sector, such as IUCN's criteria for protected areas (OPIC 1998).

Second, the industries could easily combat their pariah stereotype by agreeing to respect the right of the affected public—and foremost local communities—to say no to exploration and drilling. The public increasingly views the industries as thwarting democracy and human rights by wielding their enormous power in order to operate wherever—and however—they wish. The public considers its inability to control what happens in its own backyard, whether in Colombia or Utah, as a serious threat to the fundamental right of self-determination.

If the extractive industries agreed to adhere to a standard methodology for working with the local public that was transparent and perceived as fair, they could greatly enhance their public image. When a local community senses ownership in a project and supports exploration and development in its area, project sponsors will operate with a measure of legitimacy and avoid the degree of public ire that surrounds so many extractive projects today. Yet the converse scenario, operating without the public's approval and partnership, will undoubtedly place the company at the center of controversy.

Third, the extractive industries should demonstrate that they follow the highest applicable environmental and social standards across the board, regardless of a given host country's regulations and enforcement capacity. While adhering to high standards is simply good business in that it leads to greater efficiency and reduces the costs of impact mitigation, it could also help to sway the public perception that the sector is careless and destructive. In order to gain public credibility in this way, the industries would have to open themselves up to a greater measure of public scrutiny and oversight.

The current move toward voluntary standards, while instituted to boost the sector's image, has received only limited public support. The problem with voluntary standards is that the initiative lacks public accountability: There are no means for external evaluation, verification, and sanctioning. The extractive industries should seek innovative ways to grant the public access to their operations, so they can then demonstrate—instead of merely claim—that they are responsible actors.

Given the potential impacts of extractive industries, any industry initiative—no matter how commendable—must be coupled with an increase, as opposed to a decrease, in regulatory oversight on the part of governments and international financial institutions. While enhancing official oversight is not the responsibility of industry, it must be noted here that the international regulatory regime has thus far failed to uphold the public interest in the extractive sector and must be strengthened. In this context, just as the extractive industries expand their operations throughout the world, it will be critical to enhance official oversight of the extractive industries.

Finally, the extractive industries could improve their public image by better choosing the governments with which they collaborate. One of the main public concerns about the industries has been the perception of their collusion with repressive governments. The most high-profile example is that of Shell, which came under serious public attack in 1995 following the Nigerian government's execution of Ken Saro-Wiwa and other activists from the Ogoni People, who were opposing Shell's proposed development in Ogoniland. Shell's experience in Nigeria illustrates how a company's image can be deeply affected by the actions of its host government partners. While Shell tried to argue that it could not—and should not—be held responsible for decisions made by the Nigerian judiciary, the public nevertheless implicated the company, claiming that Shell's power and influence in the country perpetrated—and likewise could have prevented—the death of the Ogoni activists.

Some argue that Shell had a point—that its role in society is to conduct business and not become mired in a country's domestic politics. But this argument was clearly rejected by the international community, and Shell has thus suffered major damages to its public image. In some cases, despite the perceived financial gain from operating in countries governed by military dictatorships, the indirect costs of becoming the target of an angry public may be much greater in the long term.

The Future of Public Pressure

The lessons of today's public responses to the impacts of the extractive industries suggest a future in which locally affected peoples and NGOs increasingly join forces to campaign against destructive development. The public is beginning to approach the problems of the extractive industries more holistically—linking local concerns to international efforts around conservation and to human rights, labor, and poverty alleviation campaigns. Using this approach, affected communities around the world are joining together to assist each other with political support and resources, as illustrated by the cases of Mitsubishi and the U'wa.

In previous times, it was less likely for local communities to be able to successfully contest development—to prevent projects from happening in the first place. But today the message of the U'wa—demanding the right to say no—resonates throughout the world. Due to the political space that has been created through past campaigns against large-scale development and through the sophistication and collaboration of public interest groups worldwide, it is not inconceivable for the public to push successfully for the unconditional conservation of land and resources.

The empowerment of local communities may ultimately benefit the industries as well. In some cases, public empowerment may diffuse the antagonism that stems from the public's powerlessness in the face of extractive development. In turn, this may lead the public to considering the notion of project reform—the prevention and mitigation of negative impacts—instead of seeking altogether to stop a project they cannot control. In some cases, local communities may even welcome extractive development, if they believe that benefits will accrue to their region and people.

The days when the extractive industries operated largely outside the public spotlight are over. The public's capacity to monitor and evaluate development projects has increased over the last decade, and affected communities can better assess whether a project will bring them greater benefits than costs. In the last five years, the increase in the number of NGOs targeting extractive industries suggests that a proactive approach by the industries is now necessary.

If the extractive industries fail to make a true effort to understand and address the public's growing fear and mistrust, they will continue down the path toward pariah status. The challenge ahead lies in forming a meaningful partnership among all actors and stakeholders in the development process. If the industries refuse to listen to the public— and act accordingly—they will only reinforce their emerging negative stereotype with each new hot spot. But if the industries seek to involve the public as an equal player, they will have an easier path to profits, and the environments and cultures they affect will suffer less damage.

REFERENCES

Ashaye, Yemi. July 1997. *Friends of the Earth Review of Environmental Assessment for Sierra Rutile Limited.* Friends of the Earth, Washington, D.C.

Buffet, Sandy, John Ganzi, and Frances Seymour. 1998. *Leverage for the Environment: Strategic Mapping of the Private Financial Services Industry International,* I.4-I.6. World Resources Institute, Washington, D.C.

Center for Business Intelligence. 1997. International mining finance: Creative strategies, innovative sources and new partnerships for financing mining projects in emerging markets. June 1997. Toronto, Canada.

Chamberlain, Christopher. 1997. *Mining and Development: The Private Sector and the Competing Interests of the World Bank Group.* Bank Information Center, Washington, D.C.

Control Risks Group. 1997. *No Hiding Place: Business and the Politics of Pressure.* Control Risks Group, London. Pp. 3–8, 27–45, 55–60.

Da Rosa, Carlos D., Mineral Policy Center. 1998. Letter to authors.

Dasgupta, Susmita, Benoit Laplante, and Nlandu Mamingi. 1998. *Pollution and Capital Markets in Developing Countries.* World Bank, Washington, D.C.

Economist. 1996. King Solomon's mines: Mining in Africa. *Economist* 339.

International Finance Corporation. *1997 Annual Report.* World Bank Group, Washington, D.C.

Johnston, Barbara, and Terence Turner. 1998. *The Pehuenche, the World Bank Group and ENDESA S.A.; Violations of Human Rights in the Pangue and Ralco Dam Projects on the Bio-Bio River, Chile.* Report of the Committee for Human Rights, American Anthropological Association, Arlington, VA.

Kamara, Sullay. 1997. *Mined Out: The Environmental and Social Implications of Development Finance to Rutile Mining in Sierra Leone.* Friends of the Earth, London, England.

Langub, Jayl. 1998. Some aspects of life of the Penan. Paper presented at the Seminar Warisan Budaya Orang Ulu Di Miri Pada, June, Sarawak, Malaysia.

Overseas Private Investment Corporation, United States International Development Cooperation Agency. 1998. *Requests for Comments*

    on Draft Environmental Handbook; Notice Part VI 63FR 9696.
    Federal Register, vol. 63, no. 37.

Project Underground. 1998. *Drillbits and Tailings* 3 (1–6). Accessed at Rainforest Action Network, www.ran.org/ran/info_center/aa/aa117.html.

Shell Group. Shell Group, www.shell.com/f/fl.html.

World Bank. 1997. *1997 Annual Report*. World Bank Group, Washington, D.C.

————. 1998. Environmental assessment of mining projects. In *Environmental Assessment Sourcebook Update 22*. World Bank, Washington, D.C.

PART II

Oil and Gas Development
Meet Conservation

4 Reinventing the Well

Approaches to Minimizing the
Environmental and Social Impact
of Oil Development in the Tropics

Amy B. Rosenfeld
Debra Gordon
Marianne Guerin-McManus

In a continuing search for new reservoirs of oil and gas to meet grow-ing global energy demand, oil companies are expanding their explora-tion and production activities into some of the planet's most sensitive and remote ecosystems. The geographical focus of much of this new activity is the humid tropics—an area expected to be the site of a significant percentage of all new oil development in the next decade. In many cases, these tropical areas are not just host to huge reserves of oil but also are of global significance for conservation due to their rich biological diversity.

The history of oil development in sensitive tropical ecosystems has been marked by conflicts with environmentalists and isolated indige-nous communities. In many cases, this development has resulted in irreversible environmental damage and severe social disruption. New technologies and innovative management approaches—as well as im-proved communications among companies, governments, and local groups—have made it more likely that the risks of oil development in certain cases can be reduced. Petroleum development in these ar-eas will often occur in a context of weak regulatory structures, requir-ing interested companies to voluntarily establish higher standards to address the environmental and social impacts of their investments.

This chapter provides a brief overview of trends in oil development in the tropics, followed by a summary of the potential environmental and social impacts of these activities and a discussion of a range of approaches and mechanisms—principally environmental but also so-cial and financial—that can be utilized to reduce these impacts. It concludes with a series of general recommendations of ways for conservationists, governments, oil companies, development agencies, and other interested stakeholders to address these issues.

Movement into the Humid Tropics

The expansion of oil and gas development in the tropics has been driven mainly by investment from large international companies seeking new resources and attracted by the more open political and economic climate of many developing countries. The economics of oil development dictate that it generally takes a significant find to make a reserve financially attractive. Unlike many other natural resources, such as timber or minerals, which can be extracted in small amounts by individuals or local cooperatives, oil development requires large inputs of capital and equipment (UNEP and IPIECA 1995). Thus, international oil companies, like any large investor, seek economic and political stability for their investments.

Throughout many parts of the tropics, widespread liberalization of markets, privatization of oil and gas production, economic and contractual incentives for investment, and increasing political stability are making it easier and more attractive for companies to access these areas. In many cases, companies are returning to areas they once worked in, after many years of being shut out by civil unrest, political instability, nationalized oil industries, and other political, procedural, and economic obstacles to investment. Governments are also actively seeking such investment because oil revenues play an important role in developing economies and, in some countries, account for a significant percentage of gross domestic product.

Potential Environmental and Social Impacts of Oil Development

While the financial benefits of this new wave of investment for companies and developing economies can be enormous, the environmental and social impacts of oil and gas development on sensitive tropical ecosystems can be just as significant. In many cases, these impacts have resulted from a lack of understanding about the rain forest environment and the use of conventional technology and practices that, while effective in certain environments, have proven ill-suited to the fragile ecosystems of the humid tropics.

Environmental Impacts

The direct, and most obvious, impacts of an oil operation result from land-clearing and waste production. The removal of vegetation and topsoil can lead to erosion and subsequent water contamination and sedimentation of streams, especially during rainy seasons. Extensive land-clearing that removes forest canopy can expose delicate rain forest soils to increased sunlight, leading to desiccation of topsoil and deterioration of ecological integrity. Construction and land-clearing can also disrupt watershed drainage, increase the risk of flooding, and disrupt habitats and migration paths.

Improper handling and discharge of waste and toxic substances during drilling and production can also cause environmental impacts. Groundwater is particularly susceptible to contamination, notably from formation water, which occurs naturally below the surface, is extracted along with oil from the well, and can contain toxic chemicals. If the formation water is discharged into local waterways rather than reinjected into the ground, the resulting contamination can lead to serious impacts on local residents, animals, and vegetation. In addition, waste pits, when subjected to heavy tropical rains and floods, can overflow and lead to water and soil contamination. If a pit is unlined or is not landfilled when production stops, oil and other toxins can also seep into groundwater through the earthen walls.

While these direct environmental impacts can be severe, most of them are understood and can be addressed through existing technologies. However, for oil projects in remote tropical locations, it is perhaps more important to be aware of the indirect, and less obvious, impacts of oil operations. In particular, newly opened access via oil roads or pipeline paths can lead to spontaneous colonization and deforestation and habitat conversion as colonists move into previously unsettled areas, burning the forest to create homesteads and raise cattle and crops.

It has been estimated that for every one kilometer of new road built through a forested area in the tropics, roughly four hundred to twenty-four hundred hectares are deforested and colonized (Ledec 1990). Yet, despite the lushness of a tropical rain forest, its soil is actually very infertile and incapable of supporting crops for long. As soils wear out after a season or two, colonists move on to the next block of land, clearing more forest and pushing settlement further into fragile and marginally productive lands. In 1990, it was estimated that more than five hundred kilometers of oil roads in Ecuador had led to the colonization of about one million hectares of tropical forest and widespread disturbance of the traditional lifestyles of eight indigenous groups (Epstein 1995). Similarly, in the Maya Biosphere Reserve, a region under great pressure from colonization in northern Guatemala, oil field expansion and pipeline construction within and around a national park have contributed to the overall trend of an influx of colonists and expansion of slash-and-burn agriculture within the park (Sader 1996).

Social Impacts

While many environmental impacts are understood and are being addressed, social issues are still a major challenge, often because project managers have little expertise on the subject. In addition, social impacts can vary greatly, depending on a local community's degree of previous contact, integration into local economies, and proximity to existing populations.

Social impacts at oil operation sites often result from mishandled contact with indigenous groups, some of whom have had little or no exposure to the outside world. Impacts can include displacement of indigenous people as a result of increased colonization along access routes, the disruption of traditional production systems and social structures, and crowding on traditional lands. Increased dependence on outside aid can also undermine long-established societal practices. Contact with oil workers and migrants can lead to the rapid spread of diseases to which indigenous communities have no immunity.

Even in integrated towns and communities, local economies and traditional lands and production systems can be deeply affected by the presence of an oil operation. Compensation and employment schemes that are not carefully tailored to local communities can lead to dislocation, an increased workload on women, severe local inflation, and the division or destruction of family units. However, not all social impacts and resulting cultural changes are necessarily bad. Many communities have benefited greatly from jobs, better sanitation and health care, and other improvements in their standards of living.

Approaches to Minimizing the Impacts of Oil and Gas Development

To avoid repeating the mistakes of the past, the oil industry—and local and national governments—should adopt a new approach to oil development in the tropics, rethinking conventional operations and incorporating a new set of ecological and social parameters into the planning process. These environmental and social criteria should be integrated into all levels and phases of an operation—from design to abandonment—and considered as part of the cost of doing business in sensitive environments. By adopting techniques and management practices developed or adapted for the rain forest, companies can greatly reduce the footprint of their operations in these ecosystems. At the same time, nongovernmental organizations (NGOs), local community groups, and local, regional, and national governments must understand the issues surrounding these projects and work with the companies and local groups to ensure that, where oil exploration and development is appropriate, it is implemented under rigorous environmental and social guidelines and regulations.

Addressing Environmental Impact

Environmental impact assessments (EIAs) are a familiar part of oil operations today. Nearly every government requires an operating company to complete some sort of EIA before proceeding. However, in highly sensitive environments such as tropical forests, merely complying with a procedural requirement is usually not enough. In countries with less developed environmental, legislative, and bureaucratic structures, the existing legal requirements and capacity for enforcement are often inadequate to provide for full and proactive prevention of environmental damages. Operations in sensitive environments also demand a widening of the traditional scope of the EIA, considering not just the potential direct impacts at the project site but also the possibility of indirect impacts at the ecosystem level (Walsh 1998).

The EIA process should be considered not just as a one-time examination of the relevant environmental issues but rather as a tool for use throughout the operation. The EIA's predictions of potential impacts and conflicts can be used as the basis for developing long-term operational procedures and plans for environmental protection. During completion of the EIA, which should be accompanied by an equally thorough social assessment, it is important to establish a consultation process with all relevant stakeholders, including local and provincial government agencies, affected communities, and local and international NGOs. EIAs should also be subject to an outside, independent review.

Environmental monitoring and evaluation. The most thorough environmental impact assessments and management plans will not necessarily provide adequate protection against unforeseen impacts of activities in a sensitive environment. An effective monitoring program, which begins well before exploration starts, considers not only direct operations but also associated impacts to the surrounding ecosystems. Monitoring is a tool to measure the success of environmental programs and to prevent mistakes by gathering baseline information about ecosystems and measuring changes over time, allowing for an adaptive management structure that responds to changing impacts and needs.

Seismic exploration. Newer techniques and practices have been developed to greatly minimize the damage that has been associated with seismic operations in the past. Technology can improve the accuracy, resolution, and detail of seismic surveys, helping to reduce both the environmental impacts and economic costs of a surveying operation by preventing or minimizing dry holes (exploratory wells that do not reveal an exploitable reservoir of oil or gas). Seismic lines should be cut by hand, minimizing the amount of vegetation cut, leaving small, low vegetation, root stocks, and topsoil in place, and revegetating when necessary. Cutting trees wider than twenty centimeters in diameter at breast height should be avoided.

Helicopters. The expense and impact of seismic surveys and other exploration activities can be greatly reduced by using helicopters as a means to ferry people and equipment, eliminating the need to build and maintain often long roads to remote drill sites. Although helicopters are expensive to maintain, the lack of materials available for road-building, the expense of constructing a road through undeveloped terrain, and the additional mitigation and access control costs can outweigh the transportation cost differences. In addition to being fairly remote, rain forest sites may also have very unstable ground and large amounts of rainfall, making roads expensive to maintain.

Helipads should be placed as far apart as possible and, when feasible, placed in existing clearings or areas of secondary growth, away from critical habitats and breeding areas. When practical, local materials should be used to build the helipads. Outfitting helicopters with long lead lines for carrying and dropping cargo further minimizes the potential disturbance to wildlife and trees by allowing the helicopters to fly further above the forest canopy (IAGC 1992).

Personnel. In order to ensure the effective implementation of environmental technologies and management practices, workers and contractors should be informed about the potential impacts of their activities, trained in the use of new methods and technologies, and monitored to ensure proper implementation. Workers should be educated on company guidelines and required to sign contracts that allow for penalties or dismissal if these rules are broken. Having a management team that understands and will enforce the rules and regulations is important to ensure worker compliance. In a traditional oil operation, there are few people who are involved in the process from start to finish, and environmental experts are rarely called in before the operation and development stages of the project. This approach can result in inconsistent implementation and management practices. To minimize environmental impacts, it is important to integrate environmental experts into project management from the earliest stages of planning. In addition, environmental guidelines should be strictly detailed in any agreements with subcontractors during all stages of a project.

Land-clearing. Minimizing land-clearing at all stages of an operation will help prevent soil erosion and sedimentation of waterways. Bulldozer use should be minimized or avoided, but when it is necessary, operators should stay away from water and steep slopes, keeping the bulldozer blade a reasonable distance above the ground at all times to minimize topsoil disturbance. When clearing land, all trees surrounding cleared areas should be left alive as sources of seeds and organic materials, and large trees, which provide shade for the camp, should be left standing. Conducting major engineering activities during the dry season will also reduce erosion. In order to avoid permanent sedimentation or disruption of habitat, streams should be forded only when necessary, in appropriate places, and at a right angle to the stream whenever practicable.

Roads. As a general rule, oil companies should avoid building roads unless there is a real, rather than perceived, need. Thus, roads should be considered only when production operations are being established and rarely, if ever, during exploration. Instead, existing roads, riverways, and helicopters should be used whenever possible.

To prevent colonization, any new roads or points of access to the site should be strictly policed to prohibit access by unauthorized personnel. Companies should consider operating workers' base camps like offshore platforms, with all personnel, food, and equipment imported by air. Controlling access is critical for avoiding indirect impacts and requires the full support of national and local governments and local communities. A national government may have political reasons for wanting to open and maintain access to undeveloped areas, such as a desire to populate remote border regions of its country. Road-building may also involve a trade-off with local communities, who may welcome a road because it facilitates access to their homes or may oppose a road for the same reasons. Thus, it is important to understand the opinions and desires of the local people and governments and to clearly specify provisions for policing and postoperation destruction of a road in any contract documents.

Drilling. One important application of an old technology in a new setting is the use in onshore operations of directional drilling from cluster platforms, a model first used on offshore rigs. Cluster, or multilateral, wells involve drilling several wells at various angles out from a single platform. While it may initially cost more to build cluster platforms, they allow for economies of scale—with several different wells sharing the same pipelines, roads, equipment, and processing facilities—and greatly minimize the environmental impact of a drill site.

Waste. Reducing waste at all stages of an operation is far more cost-effective, and environmentally benign, than attempting to mitigate and clean up after the fact. In terms of bulk, the biggest waste disposal challenge in a remote environment is formation water. Although discharging formation water onto the surface is the easiest and cheapest method of disposal, it is also the most environmentally destructive and can lead to liability and remediation costs. To minimize both environmental impacts and financial costs, rain forest oil operations should return all produced water back into formations.

The traditional practice of digging huge pits, or sumps, to collect and hold waste is also ill suited to the delicate tropical environment because of both the danger of toxic chemicals leaking into groundwater and the need for additional clearing of large areas of land. Instead of waste pits, a large capacity tank can be used to hold liquid and solid wastes during operations. These tanks are routinely used in offshore operations and can be ferried into remote forest locations via helicopter. At the end of operations, all produced oil and water should be separated, with the water reinjected into the well and the oil processed or shipped to an off-site facility. Tanks with open tops should be screened at all times to ensure that birds and other wildlife do not become trapped in them.

A closed-loop mud system (CLMS) can also be used to manage drilling muds without a waste pit (Longwell and Akers 1992). This system employs a series of aboveground steel tanks loaded on flatbed trucks to store, process, and recycle drilling mud, cuttings, and other fluids. Although a CLMS requires higher up-front costs than traditional waste pits, the financial and environmental benefits of the system can greatly outweigh these costs. By eliminating the need to construct a several-acre waste pit, CLMS reduces the footprint of the drill site, reducing both environmental impacts and labor, materials, and equipment costs. CLMS reduces the volume of mud required and allows that mud to be recycled. In addition, the tank system can be moved easily from well to well as drilling progresses. Finally, at the end of operations, there is no need for costly reclamation of waste pits and no risk of liability from the impacts of these pits.

Daily activities and fuel use by workers at a site can also have a significant impact. Storing fuel and oil on flat, stable land, out of flood zones and above the high water mark, can help prevent pollution from spills. Storage areas should be equipped with secondary containment systems and located downslope from camps and away from other combustible materials. All storage areas should be clearly marked and routinely inspected. Vehicle maintenance standards should include similar environmental precautions. Septic tanks should be properly constructed and, when practicable, digging should be done with handheld tools rather than heavy equipment. Solid waste, such as kitchen waste, should be separated and composted at some distance from the camp in a covered hole. Nonbiodegradable waste can be collected and stored to be transported later to suitable disposal sites.

Pipelines. The debate over whether or not to build a road and the effects of increased access during oil operations extends to pipeline construction. There is disagreement over whether it is necessary to build and maintain a full road throughout pipeline operations, but it is generally acknowledged that at least some sort of swath needs to be cut for pipeline construction and maintenance. Although it is difficult, if not impossible, for vehicles to travel down a narrow path, settlers can still walk along the pipeline route into undeveloped areas, even if a trail is only a few feet wide. Thus, as with roads, it is vital to control access along pipeline paths to any undeveloped areas. Whenever possible, pipelines should be built along existing roads or clearings, minimizing new pathways. All pipelines should be constructed of double-wall piping, with pressure monitoring and automatic cutoff valves installed along their entire length, particularly on each side of a water crossing and in coastal areas (Winslow 1996).

Reclamation and mitigation. It is far more cost-effective to practice reclamation during all stages of the operation, with workers cleaning up as they go along. A company should develop a full reclamation plan for each phase of the operation before any activity is begun. Among the activities that should be included in this plan are the restoration of natural drainage patterns, the removal of temporary stream crossings, the stabilization of stream banks, and revegetation. To minimize the cost, personnel, and equipment involved in reclamation, all operations should control erosion and promote rehabilitation by minimizing the amount of topsoil removed, re-covering bare spots with dead vegetation and topsoil that have been set aside during operations, sowing grass, planting shrubs, and avoiding the use of bulldozers whenever possible.

When operations are finished at each stage of a project, all materials and facilities should be broken up and removed. All wastes should be treated and transported to an acceptable waste disposal facility. Lined waste pits should be properly sealed and the surrounding area inspected for stray waste. Compacted soils should be loosened and erosion prevention measures implemented. All roads should be rendered unpassable and the soil broken up to encourage regeneration. After removal of all equipment, a program of revegetation should be undertaken with only indigenous species, according to natural forest regeneration strategies.

Contaminated soils can be treated through bioremediation, a process in which naturally occurring microorganisms are introduced into the soil, where they break down the hydrocarbon molecules into simple carbon and oxygen compounds. In many instances, the bioremediation process results in richer soils (Shaw et al. 1995; Keeler 1991). Before leaving the site, and several times in the next months, soil and water samples from the surrounding areas should be analyzed to detect any remaining contaminants.

Addressing Social Impacts

The expansion of oil development into sensitive and remote ecosystems impacts not only the biodiversity in these areas but also their human populations, from isolated indigenous groups to communities of farmers and colonists. While many technological and management practices can reduce the environmental footprint of modern oil operations, they rarely address the social impacts. Nevertheless, as oil development moves into areas with relatively underdeveloped physical infrastructure and social services, companies are increasingly faced with the need to make decisions about and develop strategies for social issues, from health care to employment to compensation for land use. It is beneficial in the long run to study and address these issues from the earliest stages of exploration, even though a project may never go to production. If local stakeholders are not consulted and involved from the start, unexpected social conflicts at later stages may delay or even halt an operation.

Social impact assessments. A social impact assessment (SIA) should be conducted, simultaneously with the traditional EIA. The SIA process should require substantial interaction with nontraditional partners, including local, national, and international NGOs, indigenous networks, and other local community stakeholders, as well as anthropologists and other cultural experts. By providing a clear picture of the potential for adverse social impacts, the SIA is a tool for developing a successful social management plan. In some cases, early social profiles may indicate that the potential impacts or conflicts with indigenous populations are too great for a company to continue pursuing development of a particular block.

Social monitoring and evaluation. Just as it is important to monitor ecological changes through environmental indicators, it is equally vital to track social impacts. A social monitoring program, which should be underway before exploratory operations begin, uses both quantitative and qualitative tools to assess the impacts of project activities on economic and cultural systems.

Stakeholder participation. The success of any social program depends on the participation of local communities. Each project should have a full-time community liaison who is involved through the entire lifetime of the project. This community relations officer can work closely with project managers to explain the operation and its potential impacts to the affected communities and to determine their expectations, including perceived benefits from any compensation schemes. Local stakeholders should be able to review and comment on important documents, such as impact assessments and management plans.

Contact. From a social perspective, the most significant change that an oil operation brings to a remote area is the presence of hundreds of workers. Companies should be prepared with contingency plans that describe procedures for different contact scenarios between oil workers and local people. Contact between oil workers and local communities should be based on the communities' understanding of the potential impacts of contact and their informed consent to interaction. All workers should be formally trained on the population of the area, the potential for negative impacts from contact, and reporting and emergency procedures. In some cases, companies may consider withdrawing completely from areas inhabited by voluntarily isolated groups of indigenous people.

Health. Requiring workers to undergo thorough and regular medical checkups will help lessen the possible spread of disease from oil workers to local people. Complete medical records and histories of all workers should be kept on file, and vaccinations should be kept on site. In addition, medical staff should be on call throughout the entire length of the operation to monitor the health of any workers; workers who are sick should not be permitted to enter the field. If a medical emergency does arise, it is important that special programs, including the provision of food and medicines, be set up to assist the affected families.

Compensation. One of the greatest challenges of oil operations in relatively undeveloped areas is ensuring that social programs provide an appropriate level of assistance and compensation based on the community in question. The only way to avoid impacting local economies entirely is to completely isolate all project activities from local economic activity. However, this is often an impossible, and in many cases undesirable, option. Thus, all compensation schemes should be carefully tailored to the specific community based on past levels of contact, the extent of integration into local economies, and informed decisions by community members. In many communities, relying on just one individual or representative to disburse rewards equitably to all members may be ineffective and lead to conflicts within the community. In these cases, a preferred model for companies to follow may be community-based compensation, whereby remuneration is done communally and not on an individual basis.

Financial and Legal Mechanisms to Promote Environmental and Social Sustainability

The best environmental and social management practices will mean little if they are not rigorously implemented, monitored, and enforced. The potential for extensive environmental and social damage from oil development is particularly strong in developing countries with undeveloped regulatory frameworks and insufficient government capacity for implementation and enforcement. In order to ensure the conservation of biodiversity and the well-being of their citizens, developing countries need to implement progressive legislation that is supported by strong institutional capacity and innovative funding mechanisms. However, when such legislation or regulatory requirements are nonexistent or poorly implemented and enforced, it will frequently fall to an operating company to voluntarily ensure that its project complies with international standards for environmental and social management.

Legislation. National governments should enact comprehensive legislative frameworks of environmental and social regulations, from constitutional provisions and national environmental policies to more specific sectoral legislation and agency rules and regulations. In addition, individual oil contracts should supplement statutory requirements, reaffirm the company's obligation to comply with all applicable regulations, and require companies to use the best available technologies and management practices. Oil development contracts should also fill gaps in legislation and set additional requirements for particular projects in especially sensitive areas.

Environmental insurance policies. A company should be able to demonstrate to governments and citizens or indigenous groups that its insurance policy specifically covers the risk of environmental or social damages. Coverage should last for the duration of the project and for a sufficient period of time after the project's completion to ensure that any remaining problems will be addressed. A comprehensive general liability policy, while a necessary part of any oil development project, is not a substitute for an explicit pollution liability policy.

Performance bonds. To further ensure availability of funds in case of unexpected environmental or social damages, governments should require companies to post performance bonds. In some jurisdictions, a performance bond is required before any permits are granted.[1] However, even in countries where this is not a mandated requirement or where the required sum is too low, a performance bond is an important form of insurance against the potential costs of unexpected adverse environmental or social impacts. To ensure that the bond contains sufficient funds to finance any cleanup, bond amounts can be set to compensate for worst-case scenarios. If no pollution occurs, the bond will eventually be canceled, thus creating an incentive to minimize damage. Performance bonds are a useful supplement to insurance policies because the bond is automatically released in the event that environmental problems occur, while insurance companies can exercise considerable discretion over what is covered under their policies and may prefer litigation to immediate remedial action.

Mitigation trust funds. Traditional environmental trust funds function as endowment, sinking, or revolving funds, disbursing money for conservation purposes such as training environmental professionals, surveying and preserving resources, and providing environmental education (Rubin et al. 1994). Today, trust funds are increasingly being used for mitigation purposes as well. In an oil exploitation setting, if no environmental or social harm occurs, the fund may remain a traditional endowment, and the interest can be applied to basic conservation, social programs, or damage prevention measures. However, if damages do occur, the capital could be accessed and disbursed for mitigation. Money would be allocated according to the severity of the damage, and a time frame could be established to control the disbursement of money from the fund. To prevent the fund from being totally depleted in such a case, the trust would require additional contributions each time the initial principal is reduced by a specified percentage. Thus, as money from the fund is used to mitigate damages caused by oil activities, the company would supplement the remaining principal to ensure that there are still sufficient resources to correct any future environmental and social damages.

Tax incentives. Governments can also use tax incentives as a way to further conservation. For example, governments might offer tax reductions to those companies that are proactively investing in best environmental technologies and management practices that are not required by law, or to companies that are managing their lands for conservation purposes beyond their legal obligations to do so.

Conclusions and Recommendations

Conservation and oil and gas development are often in conflict within sensitive tropical ecosystems. The increasing dependency of developing countries' economies on oil revenues, coupled with the continuing global dependency on fossil fuels as an energy source, means that oil exploration in the tropics is a real and pressing problem. Thus, conservationists, local communities, and development agencies should join the debate over the future use of critical natural ecosystems, standing ready to both develop and deploy a wide range of new legal, financial, economic, and scientific tools and concepts to help inform this debate. Governments must also listen to and participate in this debate, in order to ensure that their environmental and cultural resources are not sacrificed at the expense of short-term development.

For oil companies, continued access to reservoirs in tropical countries may depend in part on public and private perceptions that they are committed to good environmental and social practices. Increasingly, companies have to answer to concerned stockholders, customers, and even financial institutions—many of whom are beginning to place more emphasis on environmental performance. Even if most companies are operating with strong records, it only takes one prominent oil spill or a major conflict with a local community to impact the reputation of the entire industry for a period of time.

A series of recommendations follows for the issues that governments, companies, conservationists, and other local stakeholders should consider when assessing operations in sensitive tropical ecosystems.

Integration of Environmental and Social Considerations throughout the Project Cycle

In order to operate effectively in unfamiliar environments such as tropical forests, companies should adopt new approaches to planning that are suited to sensitive ecosystems. Doing so will involve integrating environmental and social criteria into all stages of the project cycle. To achieve this goal, environmental and social managers should occupy senior positions on project management teams and should be consistently involved in the design and implementation of each phase of development, from planning through to remediation. The project management process should also be flexible enough to ensure that any environmental or social concerns, including unexpected ones, can be addressed.

No Development in Protected Areas

Despite the existing possibilities for impact mitigation, there are, nevertheless, certain situations in which oil development should not proceed, because the environmental and social costs will simply be too great. Our scientific understanding of biological diversity is in its infancy, and we are just now beginning to understand the importance of intact ecosystems for watershed protection, climatic stabilization, and other functions. For these reasons, oil development and other resource extraction or infrastructure development projects should not occur in national parks and other similarly protected areas. At the same time, creation of and funding for new protected areas should be an integral part of any development planning. This strategy should include increased investment in the effective management of existing protected areas, the establishment of new areas to provide maximum representation of the region's biological wealth, and the establishment of large-scale corridors to ensure long-term viability and gene flow among existing areas.

Caution in Potential Contact with Isolated Communities

Companies and governments should use extreme caution when deciding whether exploration may proceed in an area that is home to isolated indigenous communities. In some instances, the effects of contact and unplanned development on such communities can be devastating to their cultural integrity, and even deadly.

Explicit National Regulations

National governments should adopt specific standards governing oil activity on their land. These regulations should be based on internationally accepted standards for technology and management practices, and governments should ensure that the relevant ministries and oversight agencies have the capacity to effectively implement and enforce these standards and regulations. In addition, conservationists and governments alike can and should support broader use of innovative legal and financial tools to strengthen the environmental performance of private operators.

Comprehensive Regional and National Planning

Governments should work closely with scientists, environmental and social experts, local communities, and industry representatives to determine where and how oil development should proceed and where it should be prohibited. Conservationists should cooperate with such efforts to ensure that the next generation of large-scale zoning and regional planning efforts includes explicit efforts to steer roads and other development away from areas of greatest importance for biodiversity conservation.

Coordination among Companies

The expansion of oil development into relatively undeveloped areas means that several companies may find themselves simultaneously confronted with the same unfamiliar environmental and social challenges. To ensure consistency and success, companies should consider coordinating with each other, particularly in the implementation of social programs. Sharing of baseline ecological information and cooperation in the creation of compensation and employment schemes, or the setting of salary levels and medical standards, will not only be financially beneficial but will help to minimize broad ecosystem-wide environmental damages or serious conflict and misunderstanding with local communities.

Thorough Preliminary Assessment of Potential Impacts

Early evaluation of the potential environmental and social impacts of a proposed project can allow project planners to determine where and how to best mitigate these impacts. This process should move beyond simple compliance with regulatory requirements to create a tool for proactive prediction and prevention of impacts well beyond the initial government EIA and SIA approval phase. Such assessments should be ecosystem wide, covering the entire range of both direct and indirect potential impacts. In some cases, the expected environmental and social costs may be so significant that development should not proceed. It is important to establish such information before any exploration activities begin.

Use of "Best Practices"

Where development does proceed, companies should use the most advanced and efficient technology and management practices available. Even where this obligation may not be mandated by law, it is the responsibility of the company to utilize best international standards on a voluntary basis or to demonstrate why a specific advanced technology or practice will not work in a particular situation.

Infrastructure Siting

Careful planning and siting of project infrastructure is crucial for avoiding long-term environmental and social damage and excessive mitigation and liability costs. Determining the most vulnerable areas of a concession, from unique ecological features to the homelands of isolated indigenous groups, will help to minimize unexpected obstacles to development that might cause irreversible damage to local habitats and cultures and delay, or even halt, a project. This planning should be done as part of an environmental and social assessment phase before any project activity begins.

Ecological and Social Monitoring

Monitoring of the ecological impacts of major natural resource extraction projects should be a fundamental part of industrial planning in sensitive ecosystems. Well-developed ecological protocols can provide an important baseline for assessing performance and determining the impacts of project activities on the surrounding ecosystem, cultures, and economies. Monitoring can also determine whether programs instituted by the company are having the desired effect and can suggest alternatives if they are not.

Proactive Engagement with Stakeholders

Major development projects of any nature should take place within the context of public participation in decisionmaking. Stakeholders will include local communities, indigenous federations, regional, national and international NGOs, and local and regional governments. Early and consistent communication among the company, the national government, and these stakeholders will help to identify potential sources of conflict and allow issues raised to be addressed in a timely and comprehensive manner. This communication should include the involvement of social experts who are familiar with local cultures.

Understanding the Issues

In order to effectively represent the environment and local communities and constructively engage with international companies and national governments, NGOs and local community groups should educate themselves on the issues and trends in global oil development. These groups should have technically sophisticated knowledge and arguments to support their opinions and recommendations on siting and development of oil projects on their land or in areas in which they work. This knowledge will require not only learning technologies and methods but continually monitoring developments and trends in the industry.

Creative Conservation Opportunities

The use of "best practices" is not the only way that oil development can have a positive impact on conservation. Through direct investments in conservation—such as funding management of nearby protected areas, encouraging the creation of new protected areas, or establishing endowments for resource conservation—companies can have an impact far beyond the borders of their project sites and well past the lifetime of a particular reservoir of oil.

Acknowledgment   This chapter is based on Rosenfeld et al. (1997).

NOTE

1. Performance bonds are a standard feature in construction contracts in the United States and are increasingly being used for environmental purposes. For example, the Texas Parks and Wildlife Department requires performance bonds of $1 million for all pad sites before allowing mineral recovery on their lands. See Texas Parks and Wildlife Department (1996). Similarly, the Florida Department

of Transportation requires performance bonds for oil spill contingencies under section 377.2425 of the Florida code, although the Florida Supreme Court recently held that the Department of Transportation could not require a performance bond if the oil company had already paid an annual fee into the Florida Petroleum Exploration and Production Bond Trust Fund. *Department of Environmental Protection v. Coastal Petroleum Company*, 660 So. 2d 712 (1995).

REFERENCES

Epstein, Jack. 1995. Ecuadorians wage legal battle against US oil company. *Christian Science Monitor*, September 12, p. 10.

International Association of Geophysical Contractors (IAGC). 1992. *Environmental Guidelines for Worldwide Geophysical Operations.* International Association of Geophysical Contractors, Houston, Tex.

Keeler, Robert. 1991. Bioremediation: Healing the environment naturally. *R & D*, July, p. 34.

Ledec, George. 1990. *Minimizing Environmental Problems from Petroleum Exploration and Development in Tropical Forest Areas.* World Bank, Washington, D.C.

Longwell, H. J., III, and T. J. Akers. 1992. Economic environmental management of drilling operations. Paper IADC/SPE 23916, presented at the IADC/SPE Drilling Conference, February 18–21, New Orleans, Louisiana.

Rosenfeld, Amy B., Debra L. Gordon, and Marianne Guerin-McManus. 1997. *Reinventing the Well: Approaches to Minimizing the Environmental and Social Impact of Oil Development in the Tropics.* Conservation International, Washington, D.C.

Rubin, Steven, et al. 1994. International conservation finance: Using debt swaps and trust funds to foster conservation of biodiversity. *Journal of Political and Economic Studies*, p. 19.

Sader, Steven A. 1996. *Forest Monitoring and Satellite Change Detection Analysis of the Maya Biosphere Reserve, Petén District, Guatemala.* Conservation International and U.S. Agency for International Development, Washington, D.C.

Shaw, Buddy, Catherine S. Block, and C. Hamilton Mills. 1995. Microbes safely, effectively bioremediate oil field pits. *Oil and Gas Journal*, January 30, p. 85.

Texas Parks and Wildlife Department. 1996. *Staff Guidelines for Mineral Recovery Operations on Department Lands.* February.

United Nations Environment Programme (UNEP) and International Petroleum Industry Environmental Conservation Association (IPIECA). 1995. *Technology Cooperation and Capacity Building.* United Nations Environment Programme and International Petroleum Industry Environmental Conservation Association, London.

Walsh Environmental Scientists and Engineers, Inc. 1998. *Estudio de Impacto Ambiental: Proyecto Perforación Exploratoria Pozo Candamo—1X, Lote 78 (Tambopata).* Prepared for Mobil Exploration and Producing Perú, Lima, Peru.

Winslow, Kelly S., director of engineering, U.S. Environmental Group, Inc. 1996. Personal communication with author, November 12, 1996.

5 An Industry Perspective on Environmental and Social Issues in Oil and Gas Development

Case Studies from Indonesia and Ecuador

Robert Kratsas
Jennifer A. Parnell

For several decades, oil and gas companies have operated in some of the world's most sensitive environments, including tundra, rain forests, coral reefs, mangroves, wetlands, and deserts. Millions of dollars have been spent by oil and gas companies to improve health care, build housing and medical facilities, enhance education and job training, and improve daily life for people in nearby communities. Recent years have been marked by strongly worded policies, innovative technologies, reduced impacts, increased corporate environmental performance reporting, stakeholder consultation, and comprehensive environmental management systems.

As exploration and production companies focus their efforts in the humid tropics—long recognized as centers for biodiversity and indigenous culture—environmental and social issues become intensified. Energy development will continue to occur in sensitive environments throughout the world, and all energy production inevitably involves some impact. The challenge for energy companies in this arena is to move beyond rhetoric, beyond the successes and failures of the recent past, and establish a new level of trust, both with the communities in which the activity occurs and with the global community.

Voluntary corporate disclosure of environmental performance and the old model of village-by-village compensation (the so-called beads and trinkets approach) have faded. It is not enough to be "seen to be green" in the humid tropics; there is a new corporate accountability. Energy companies in these regions of the world will increasingly be judged not only on technical skill and economic performance but increasingly on who we are, how we behave, and what issues we support. To compete as energy providers into the next century and beyond, we must have a vision that recognizes the broadening corporate environmental and social responsibilities. These issues must be addressed if corporations are to retain their license to operate in the humid tropics. Oil and gas companies will be judged on how well we meet established performance criteria, how well we communicate with the various groups who demand information on environmental and social issues, and our readiness to engage in change and dialogue.

Two projects by Atlantic Richfield Company (ARCO)—Pagerungan and Villano—demonstrate our commitment to environmental protection and social development. They showcase new technologies—and perhaps more important, new corporate attitudes. Further integration of environmental and social objectives can be achieved through collaboration with NGOs, and innovative approaches to communication with the full range of stakeholders.

Pagerungan: Community Development on a Small Island in Indonesia

In 1985, substantial commercial gas reserves were discovered near the edge of the continental shelf in the province of East Java in Indonesia. The Pagerungan Field is located off the northern shore of Pagerungan Besar Island in the southeastern Java Sea, north of Bali Island. In the great Indonesian archipelago—seventeen thousand islands spanning three time zones in the western Pacific—Pagerungan Island is less than a dot (only 355 hectares), too small to be found on most maps. However, Pagerungan Island is a large part of ARCO's daily operations as the principal natural gas supplier to the ever-increasing Indonesian market. ARCO pipes gas each day to Surabaya, the second largest city in Indonesia, and to customers throughout East Java, where it is used to generate electricity. The gas processing plant, built in 1992–93, handles 350 million cubic feet per day of dry gas and three thousand five hundred barrels per day of condensate. Drilling of nine initial wells (and five subsequent wells) began in September 1993. A subsea pipeline twenty-eight inches in diameter was built by Pertamina in 1993, and gas and condensate sales began in 1994. ARCO's environmental and community development programs at Pagerungan Island established a balance between the project, the environment, and the community.

Pagerungan is a place of great natural beauty, with sparkling seas, brooding mangrove swamps, seagrass, and coral. There are no surface waters, streams, lakes, or ponds on Pagerungan Island, but reef flats, coral rubble, live coral, and coral fringe are very extensive. The reefs provide nursery and breeding areas for commercial and recreational fish, and the reef flats are recognized as winter feeding grounds and

Figure 5.1   Indonesia.

staging areas for shorebirds. The climate is tropical and monsoonal;
circulation and currents are controlled largely by the monsoon. Annual rainfall ranges from fifteen hundred to two thousand millime-
ters, and temperatures range from 22 to 33 degrees Celsius.

The marine ecosystem found around the project area is typical of
that found throughout the tropical water region. Shallow limestone
platforms support extensive macroalgal and seagrass communities as
well as marine life. Although some reefs exhibited moderate human
and natural degradation, generally their conditions ranged from poor
to good at the start of the project. Seawater clarity is quite high (generally greater than twelve meters), with temperatures ranging from 26
to 29 degrees Celsius. The island is rich in marine resources, with
some protected species, including seabirds, sea turtles (green and
hawksbill), and Dugongs (*Dugong dugon*).

Pagerungan is home to some four thousand Indonesians in eight
hundred households, who for centuries have made a living from fish-

ing and subsistence farming. The culture of the island is derived from age-old and strict religious traditions of the Boegenese (South Sulawesi) and Madura (Madurese), both with a strong fishing heritage. Eight out of ten islanders fish for a livelihood; the rest work at subsistence farming. Total catch averages fifteen hundred tons a year. Principal commercial fisheries harvest pelagic fish (tuna and mackerel), reef fish (grouper and snapper), milkfish (*Chanos chanos*), and ornamental fish. Due to limited refrigeration, the industry traditionally focused on dried and salted fish. Fresh fish could never survive the two-day trip to the markets of Surabaya. On land, maize, cassava, soybeans, and fruits are grown, but only coconuts are grown in sufficient quantity to allow a surplus to be sold to other islands.

Before gas production began, education and health care had been modest. Only three elementary schools and one madarasah (Islamic) school, all in need of repair, served 650 students. There were few books, writing materials, or other supplies. Existing public facilities were not in good condition. A branch of the public health service (Puskesmas Pembantu) was staffed by one certified paramedic, with occasional visits from outside professionals.

In opening Pagerungan Island to development, the Indonesian government was determined that all aspects of local life should be addressed— not simply that the gas be developed and the pipeline constructed, but that steps also be taken to protect the native population and the environment. Indonesia wanted the natural gas strike to bring positive change through careful and balanced management. ARCO agreed.

As ARCO undertook gas development in the early 1990s, we worked with local authorities and Pertamina, the state oil company, to develop a plan that called for ARCO to:

- Protect the physical environment;
- Contribute to cultural and economic well-being;
- Establish harmonious relations with the islanders and their government;
- Modernize the fishing industry; and
- Improve the educational, health, and cultural infrastructure.

The Environmental Impact Document (Amdal) for the Pagerungan gas project was prepared and approved by the Indonesian government in 1991. The preliminary environmental assessment (PIL) and Amdal identified sensitive and valuable marine resources of the area, impacts, statutory measures, procedures, and environmental mitigation measures. Potential impacts included:

- Atmospheric changes due to gas flaring, generators, air transportation, and construction;
- Increases in noise levels from flaring and generators;

- Deterioration in seawater quality from jetty and harbor construction, waste disposal, and leaking during loading of condensate;
- Deterioration in the quality of shallow groundwater in surface layers due to intrusions of waste disposal and drilling;
- Decrease in the amount of available freshwater and groundwater due to decrease in land surface area reserved for absorption of rain water;
- Damage to coral reefs from construction and drilling;
- Disturbance of marine life;
- Disturbance of nursery grounds;
- Social conflicts arising from lack of local employment, disagreements over land acquisition, deterioration of the physical environment, decreased agricultural production, and decreased fishery catches; and
- Positive impacts during construction and operation for local labor.

To achieve balance among the needs of the project, the natural environment, and the neighboring community, technical, social, and institutional mitigation options were developed. To minimize impacts on shallow groundwater and seawater quality, ARCO developed a plan to contain, treat, and remove all solid waste from the island. The company also built a two-unit sewage treatment plant with a capacity of fifty thousand gallons per unit. Treated liquid waste is now discharged into deep water beyond the reef (fifteen meters or greater depth). Treated water is tested daily to ensure Indonesia's chemical and biological water quality standards are met. Site-specific permits ensure that environmental sensitivity, ambient water quality, and present condition of discharge sites are evaluated. During drilling, ARCO also installed one temporary mechanical slurry discharge unit to treat drill cuttings before injection.

A reverse-osmosis desalination plant was installed to supply freshwater for the project (150 cubic meters per day) so that shallow groundwater would remain available. The unit also provides thirty cubic meters per day of clean water for the community. In addition, ARCO built a catchment rain area (fifteen hundred cubic meter capacity), conducted revegetation programs, and minimized land clearing. Construction activities were carried out only during daylight hours, and vegetation was left in place around the facility to provide an aesthetic and noise buffer. The height of flare stacks was raised twenty meters to dilute emissions concentrations.

During construction, the company also made efforts to reduce impacts on the coral reef and nursery grounds. However, construction of the jetty, outfall line, and facilities inevitably resulted in some damage to the reefs. The blasting and dredging of the harbor (causeway and harbor basin) caused the most significant impacts: sediment-loading and turbidity resulting from suspension of fine silt and mechanical damage to the reef caused by the cutter head and other equipment. On a large scale, this process can destroy the physical structure and ecological communities of the reef. Algal zooxathella lead symbiotic lives with corals and require sunlight for photosynthetic processes to produce the calcium carbonate skeletons necessary for coral growth.

To avoid this impact and also to minimize sediment material, the dredged soil was stockpiled in the harbor on a reef flat and later used for the roadway base and fill material. The siltation that appeared in this area was dispersed over the reef flat or diluted rapidly in the open sea. Based on underwater monitoring of the fringing "live" reef, data indicated that sedimentation rates during construction phases in the vicinity of the harbor area were consistently twice the rate measured at a control site. But the impact of increased sediment on the reef was only of slight significance and occurred during a short time period. Concentrations were only distributed within a few meters of the discharge point. A number of reef species exhibited symptoms of stress, but the sponges exhibited a high tolerance, and the reef continued to grow. Only a small reef surface was partly covered with a thin layer of sediment, and in a few days the reef was sediment free.

Six months after construction of the harbor, the reef growth was normal, with ranges between 0.5 and 1 percent every two months. Nevertheless, to restore degraded coral reefs and to provide substrate for coral growth and new habitat for fisheries, ARCO installed forty pyramid module artificial reefs in three different locations in 1992 and 1993. This artificial reef was installed three months after the dredging and blasting of the harbor. Presently, the artificial reefs are providing new fishing grounds for the islands and are covered by soft coral.

ARCO also planted seventeen hectares of *Rhizophora* mangrove forests (thirty-five thousand trees) adjacent to project facility areas to help control coastal erosion, stabilize sediments, protect adjacent coral reefs, and provide nursery and breeding grounds for commercially important fish, crustaceans, mollusks, and other marine life. These wetland forest ecosystems are expected to generate a wide range of renewable resources that can be harvested indefinitely if properly managed. In addition, the mangrove can play a valuable role in supporting fisheries, including the population of milkfish.

To support mitigation efforts, ARCO has conducted underwater surveys, coastal land suitability evaluations, and reef and mangrove growth studies. Prior to these activities, ARCO notified fishermen's groups and local authorities regarding the timing of operations, mitigation plans, and maintenance of reef and mangrove growth. As the project progressed, ARCO worked closely with the Research Centre for Tropical Biology of Indonesia to conduct marine environmental management conservation and monitoring. ARCO installed three permanent and one temporary transect line and sediment trap stations in shallow and deep water for sedimentation and macrobenthos monitoring, focusing on turbidity and siltation. Permanent and temporary stations were installed for nursery ground monitoring (*Chanos chanos*) as well as seawater quality monitoring. Studies do not show any negative impact on fish or other marine species population in the vicinity of the discharge point.

For example, collection of milkfish is the second primary income for local fisherman, and milkfish live within the littoral zone, separated from the open sea by a flat reef zone. Construction can be highly destructive to nursery grounds by altering water flow around tidal zones and causing ecological changes, such as aborting larval development and interfering with physical, particularly reproductive, processes. To avoid this impact, construction was done during the off-season for milkfish. Monitoring data show that the number of milkfish larvae hatched per unit hour and the overall status of the stock are normal.

ARCO also made efforts to manage unanticipated environmental impacts during exploration. One of the initial wells drilled in 1993 began to flow gas during shallow drilling as a result of a change in ocean pressure. The flow was diverted, and the well was quickly brought under control. There were no injuries to site personnel or local inhabitants, no significant loss of property (drilling unit), and no damage to the surrounding village. Local authorities were briefed throughout the operation, with safety as the highest priority. Impacts from the release of gas and salt water were monitored, and monitoring shows the environment fully recovered in less than a year.

Today, years of environmental monitoring and quarterly environmental inspections by the government of Indonesia show that construction and operation of the Pagerungan gas project has been achieved with minimal environmental impacts. The ecosystem remains in very good condition.

ARCO was committed to providing lasting socioeconomic benefits to the island community. During the three years that followed initial development, ARCO designed a community development program that included construction of a junior high school and three elementary schools, as well as housing for school headmasters and a paramedic. Since the population of Pagerungan is Moslem, ARCO built an Islamic school and reconstructed a mosque. In addition, the company built a village hall and sports and youth center, renovated the Puskesmas (health clinics), and increased the supply of freshwater and electricity. As part of the new program, Pagerungan got its first ice plant (five-ton capacity) and cold storage facility (twenty-ton capacity), opening new and potentially lucrative markets to local fishermen. The living standard and income of local people has improved without any social conflict involving ARCO, local residents, and government.

Approximately five hundred local workers were hired to work during project construction. After completion of the project, approximately one hundred people were hired as local labor. The community development project was designed so that those workers who could not be accepted to work in the plant could go back to their original trade, without causing any disappointment. Further programs were initiated to help local people improve their income by their own efforts, including a cooperative and a local market and improvement in education through student scholarships.

At Pagerungan, ARCO worked at many levels to achieve the following:

- Prevent the growth of negative attitudes and avoid social conflicts;
- Protect project facilities from marine environmental damage (abrasion, erosion, etc.);
- Implement marine conservation programs for protection of sensitive habitats;
- Improve technical guidance and laws for oil and gas projects;
- Stimulate employees and local people to improve knowledge and skills in environmental management and conservation; and
- Stimulate environmental growth.

As a result of ARCO's efforts, the Pagerungan project received a technical achievement letter from the government of the Republic of Indonesia in recognition of ARCO's outstanding achievement in marine environmental and conservation efforts. The Pagerungan project also received an Environmental Achievement Award from Pertamina (the state oil and gas company) for two years in a row, 1996 and 1997. Pagerungan now serves as a model for the East Java region, and lessons learned on this small island can be applied to other developments throughout the tropics.

Villano, Ecuador: Setting a New Standard East of the Andes

Rain forest covers nearly half of the land area of Ecuador, which is located on the northwest shoulder of South America's Pacific coast. The tightly knit canopy of broadleaf evergreen trees admits only traces of sunlight; the forest rolls down the eastern slope of the Andes and into the Amazon River drainage area, where ARCO is developing a major oil discovery located in Pastaza Province, near a cluster of villages known as Block 10 Villano.

Figure 5.2   Ecuador.

Villano sits in a uniquely sensitive environment. Ecuador's rain forests are considered to be among the world's most biologically diverse regions. They provide a home to perhaps 10 percent of the world's plant and animal species, including 120 mammals, 500 fish, 600 birds, at least 12,000 plants, and untold numbers of insects. The rain forest is especially vulnerable to industrial activities like oil and gas development, as has often been documented in the last half century.

The Ecuadorian rain forest is also home to 250,000 indigenous people, many of whom depend on the rain forest for their livelihood and are anxious to conserve this resource. Block 10 includes several indigenous groups, principally Quechua, Huaorani, and Shuar. Ten small villages inhabited by around three hundred Quechua-speaking people are located near the Villano well site. These people have known outsiders since the late 1930s, when American Protestant missionaries established themselves in the area.

Several oil companies, including Shell, Anglo-Ecuadorian Oil Fields, British Gas, PetroCanada, Unocal, and Petrobras have prospected in this area. Exploration activity in the Oriente Basin began in 1937 when Shell drilled two wells; Shell joined with Exxon for four more wells, which were eventually abandoned in 1950. In 1964, a Texaco/Gulf consortium was awarded a large tract north of Villano; it made several large discoveries in 1969.

Recent interest in the Oriente Basin is increasing. As of March 1998, several major companies and consortiums have received contracts from the Ecuadorian government to explore for oil and develop new discoveries in the area. In addition to ARCO/Agip in Villano, these companies include Elf Hydrocarbons, Occidental, YPF, Pucalta (a Canadian firm), and others. Meanwhile, Petroecuador (the state oil company) has taken over Texaco's former operations and produces around three hundred thousand barrels of oil per day.

Many indigenous people near Villano have worked, at least temporarily, for oil companies and are interested in expanding opportunities in the market economy. Others would like to improve the quality of their lives through assistance for education and health care and direct support. Still others remain opposed to petroleum development in any case.

Over the past twenty years, indigenous peoples in Pastaza Province have formed several federations to obtain legal title to community lands and address other issues. ARCO's neighbors in Block 10 are represented by the Organization of Indigenous Peoples of Pastaza (OPIP), the Association for Indigenous Development, Amazon Region (ASODIRA), and the Federation of Indigenous Peoples of Pastaza, Amazon Region (FIPPRA).

Oil has been a significant factor in Ecuador's economy since the 1920s; today this factor accounts for 11 percent of the country's gross domestic product, 43 percent of its export revenues, and over half of its government's income. Past environmental practices in the rain forest have been strongly criticized by native people, environmental groups, and international organizations. At Villano, ARCO set out to prove that Ecuador's rich oil resources can be developed to benefit the country at large while preserving the rain forest and the well-being of its 250,000 residents.

In July 1988, ARCO Oriente, Inc. (AOI), a subsidiary of ARCO and its partner Agip Petroleum (Ecuador) Ltd., signed a twenty-year "risk service" contract for exploration and production in Block 10. In 1992, ARCO announced a major discovery at Villano. Under its agreement with Agip, ARCO will operate this field and holds a 60 percent share. Reserves are estimated at 170 to 200 million barrels.

In March 1997, the Ecuadorian government approved the Villano development plan, including an eighty-one-mile pipeline to bring the crude to market. Under the contract, ARCO and Agip will begin to recover the costs of exploring and developing the Villano field when oil is delivered to the Trans-Ecuadorian Pipeline System (SOTE in Spanish). The partners expect to recover 171 million barrels of oil from the block. The EIA for Villano was approved in April 1998, and the project is now under construction. Operations are scheduled to begin in April 1999. The development plan has been guided by two goals: minimizing environmental impacts and maintaining an open and frank dialogue with indigenous communities.

Facilities planned for Block 10 include seven production wells and three injection wells for produced water. Peak production from the wells is expected to be thirty thousand barrels per day. A twelve and three-quarter inches in diameter flowline will run thirty-eight kilometers between the well site and the Central Processing Facility (CPF). A secondary sixteen inches in diameter pipeline will run along the lower reaches of the Andes for 140 kilometers from the CPF to a pumping station near Baeza, where it will join the SOTE.

A key factor in gaining the government's go-ahead was ARCO's development plan, which is a significant departure from the past. Technologies designed to protect fragile ecosystems in the North Sea, the Gulf of Mexico, California, and Alaska's North Slope will be adapted to achieve ARCO's goal of drilling the least number of wells possible to efficiently drain the reservoir.

The site will be treated similarly to an offshore development. No roads will be built in the area. Helicopters will be used to transport equipment, personnel, and supplies to the construction site and pipeline construction camps. Every effort will be made to minimize the footprint on vegetation from seismic lines, helipads, campsites, and drill sites. The total footprint of the well site will be only 2.7 hectares. The well site will be operated remotely from the CPF to minimize surface area impacts in the forest.

One key innovation involves using a monorail system for building and maintaining the flowline through primary forest from Villano to the CPF, which will be located outside pristine rain forest areas. The monorail will eliminate disturbance of the rain forest canopy by heavy equipment during construction and permanent disruption of the canopy by the removal of large trees for a pipeline and road right-of-way. The monorail will also serve as its own supply line. All materials, equipment (except six small tractors), supplies, fuel, and even meals will travel by monorail to construction sites. This will allow ARCO to reduce the width of the flowline path between Villano and the CPF from eight to twelve meters to less than two to five meters. The number of trees to be cut will be reduced from the conventional 500 per kilometer to 130 per kilometer, and disturbance of large mature trees will be minimized since the system can "bend" around them. Because the canopy will remain undisturbed, there will be no permanent habitat disruption or further erosion of the canopy due to wind damage.

The monorail will be built using Kaiser "walking tractors" specially designed to work in sensitive environments. These are relatively small and lightweight and can maneuver on rubber "swamp buggy" tires and extension legs over rough topography without tearing up the earth and root systems or stumps. They are also adaptable for a number of other uses, including sawing, pile driving, and drilling, and therefore replace all equipment otherwise required for construction.

The flowline will be constructed on an assembly-line process as the monorail is extended from Villano toward the CPF. Loads of pipe will be transported to small "drop sites" every three kilometers along the route by helicopter, and the pipe will then be moved through the forest by monorail and welded together. When completed, the monorail will stand about one meter tall—just high enough to accommodate the twelve-inch pipeline and a cable along which supplies and maintenance will travel.

After construction, the monorail system will be left in place for twenty years. Some of the helicopter drop sites along the monorail will remain, and materials and equipment will use the monorail to reach areas that may require repair.

Where possible, ARCO will leave topsoil undisturbed to promote restoration and revegetation. A native plant nursery will be created, and these plants will be used for reforestation of camp areas and flowline as needed. All disturbed areas will be restored as closely as possible to their original state.

The ten temporary construction camps have been carefully designed to minimize impact by building vertically rather than horizontally and utilizing reusable, lightweight materials in modules. Only steel and factory-processed wood will be used. The steel buildings will be transported unassembled by helicopter and can be handled without equipment. They will sit on adjustable legs on the ground so that no roots or stumps are disturbed.

Once the flowline leaves the forest, it travels through a heavily farmed area where the CPF is located. The route was carefully selected to avoid natural barriers such as rivers and streams and to follow existing natural paths. In addition, ARCO has purchased an area where lumber companies have already begun to cut virgin forest and will develop a buffer zone against further deforestation.

The CPF will be located in farmland accessible from Puyo by an existing twenty-kilometer road. This facility will occupy only thirty hectares on a tract of nearly 230 hectares purchased by ARCO. At the CPF, water brought to the surface with oil at Villano and pumped along the flowline will be separated and reinjected into subsurface geological formations. The oil will then travel by a sixteen-inch secondary pipeline along the lower reaches of the Andes to a point near Baeza, where it will be measured and pumped into the SOTE. The route for this pipeline lies mostly along existing road and travels through farming areas.

The pipeline will be buried, except for the four kilometers where it crosses the Antisana National Park just before Baeza. This measure is necessary because the existing roadbed is subject to landslides and may not be able to safely support the pipeline. This pipeline will follow the crests of hills to minimize erosion. ARCO is working closely with the Ecuadorian government's national parks service (INEFAN) and the Fundación Antisana (a nonprofit conservation group) to design the pipeline, develop contingency plans, and practice drills. During construction and operations, local residents will be trained to monitor the project's environmental impact from Villano to the CPF. Monitoring in the Antisana Park will be overseen by INEFAN and an internationally recognized environmental consulting firm.

Does ARCO's "offshore development" at Villano cost more than a conventional approach? The biggest difference involves bringing electric power via cable from the CPF to the well site. Other costs—for example, using helicopters and walking tractors to construct the flowline instead of building a permanent access road—will even out once the full twenty-year life cycle of road maintenance and other expenses are taken into account.

During its exploration program, ARCO was slow to recognize the importance of building partnerships with native communities and organizations. As a result, the company soon found itself involved in disputes with its neighbors in several areas. To resolve these disputes, area managers met with indigenous representatives at company headquarters in suburban Dallas in March 1994. The outcome of this meeting included a major new initiative: convening a "technical environmental committee" with participants from ARCO, Petroecuador, and three Indian organizations in Pastaza to plan and implement community development projects for Block 10 jointly over the next twenty years. Although this process has been difficult at times and is still unfolding, ARCO is committed to developing a model partnership to provide sustainable new social and economic opportunities while minimizing or eliminating potential disruption. ARCO has cooperated with community development efforts in Block 10 since 1988 and is committed to respecting indigenous territories and cultures.

The Technical Environmental Committee (CTA) continues to play a central role in these efforts. The main objective of the CTA has been to create a channel of communication between the company and indigenous organizations on environmental issues that otherwise could become an arena for continuing conflict. In retrospect, this group has provided an ongoing forum for exchange of information and indigenous involvement in environmental assessment and monitoring throughout the project. The process has been slow and painful at times, but large steps have been made toward establishing trust and creating transparency. The CTA is important, because in a region where such efforts to engage the public in decisionmaking are often challenged as wasteful and unnecessary, ARCO has taken a different view. Company officials in Ecuador believe that continuous confrontation and conflict will not allow the company to work effectively, nor would the interests of our neighbors be served. Common sense and business sense have helped forge a new attitude.

Recent events in the Villano development with regard to indigenous relations and dialogue may call into question the effectiveness of the process. Some may even believe that the process has failed. Quite to the contrary, ARCO does not believe that the process has failed. Recent indications from the indigenous people are of a strong desire to resume dialogue as quickly as possible.

What events have caused this concern? With the approval of ARCO's EIA in April 1998, the company had begun work to construct a Villano drill site and associated flowline to carry oil out of the rain forest to its CPF. Through communication with the Technical Environmental Committee, ARCO believed that agreement had been reached with the various indigenous organizations on the basic development plan, the EIA, and the rights-of-way necessary for the project to proceed. Yet on July 19, 1998, three ARCO representatives and one contract employee were taken hostage in the village of Santa Cecilia, near Villano. A few days later, three other contract workers were abducted and held in San Virgilio, another small community not far from the CPF. By July 29, the three ARCO hostages had escaped, and the remaining hostages had been released.

What caused these events? In our view, several factors played an important part. Most residents of Santa Cecilia belong to ASODIRA, a small indigenous group that was formed several years ago when members of ten Villano villages seceded from the area's main Indian federation, OPIP. In founding their own organization, these communities hoped to work directly with ARCO to secure local health, education, and economic benefits for their communities.

By July 1998, however, some of ASODIRA's leaders decided to take a tougher stance on the Villano project. One major disagreement was Santa Cecilia's demand for ARCO to construct a road from the existing interprovincial highway through virgin rain forest to the Villano communities. Such a road would be counter to ARCO's environmental vision for the project. Nonetheless, Indians near Villano face the unresolved problem of transporting their products to market and want a road. A longstanding discussion with these communities was how to provide economic alternatives without endangering the forest.

Several days after the kidnapping began, OPIP representatives and other Indian leaders from outside of Santa Cecilia met the kidnappers there. On July 26, they issued a joint communiqué demanding significantly increased participation in government petroleum policy and a share of production revenues. At this point, according to ARCO community relations manager Carlos Villarreal, who was one of the hostages, regional indigenous leaders expected the kidnappers to release him and his colleagues. But at the last moment, the kidnappers objected that their main demand—a road through the jungle—was not reflected in the broader communiqué. As tensions among the groups rose, one OPIP leader told Villarreal that his captors were planning to move them deep into the jungle. He and his colleagues escaped to a nearby army camp.

The second kidnapping, which took place a few days after the events in Santa Cecilia began, essentially represented a "copy-cat" incident involving three employees of ARCO's pipeline construction firm, Conduto. Apparently, residents of San Virgilio wanted ARCO to speed up completion of the projects specified in the right-of-way agreements and also decided to add a few things that they had neglected to include in the original agreement. With help from the provincial governor, this conflict was quickly resolved and the hostages were released.

Will these incidents have a long-term impact on ARCO's relationships with indigenous people? Paradoxically, such relationships may well be strengthened as the Villano project moves forward. In response to indigenous demands, the new ministers of energy and government have committed to opening a dialogue with native organizations on fundamental questions of regional development, improved infrastructure, and other matters. ARCO may find that it can play a supporting role in expanding opportunities for education, health care, and economic growth without taking primary responsibility for long-term social programs. ARCO can then focus on working more intensively with its immediate neighbors in Block 10. Obviously, additional measures will be required to ensure the security and safety of our people. However, these events will not affect our commitment to maintain the dialogue process with the indigenous people. We have always sought to be good neighbors and will continue to do so.

Key Themes

These projects describe a new standard for petroleum development in sensitive environments. At Villano and Pagerungan, this standard requires more than choosing to develop a resource responsibly. It involves more than designing facilities and operations to protect the environment and local people. It includes more than baseline environmental studies, integrated project design teams, mitigation, and monitoring. Recurring themes in ARCO's strategy include:

- Clear corporate environmental policies and procedures;
- An organizational structure that supports two-way communication between corporate functions and field units;
- Effective environmental compliance and information management systems;

- Top-level support and commitment to environmental protection and community development goals;
- Integration of environmental and social issues into project planning and conceptual engineering;
- Minimizing disturbance of natural surroundings;
- Detailed environmental and social assessments;
- Risk assessment and risk management, including social and economic risks;
- Independent monitoring and evaluation of environmental and social impacts; and
- Comprehensive environmental and social awareness training.

In order for this approach to succeed, companies and conservation groups alike need to focus on sustainable methods to maintain natural systems and community-based management initiatives. They must also foster trust and cooperation. Preserving natural habitats and the biodiversity they support is a start, but it is not enough to create parks and refuges and conduct surveys. Engineers must have an understanding of the environments they develop, and creative technical thinking must be encouraged. Both corporations and environmental groups must work to gain the capacity to see beyond park boundaries and develop long-term solutions to land use issues. The key is to develop methods that allow for use of the land while maintaining its ecological integrity and functions. To achieve a balance between the world of development and biodiversity in the humid tropics, there must be a shift from a culture of confrontation to a culture of dialogue. Support for and encouragement of sound environmental planning and management will help ensure that ecological integrity and functions of natural systems will be maintained when and where development occurs.

Acknowledgment The authors gratefully acknowledge extensive contributions and review by Abdul Adjiz Saleh (Atlantic Richfield Bali North), Amir Hamzah (Atlantic Richfield Bali North), Bob Alcorn Jr. (ARCO), Herb Vickers (ARCO Oriente, Inc.), Sixto Mendez (ARCO Oriente, Inc.), and Robert Wasserstrom and Susan M. Reider (the Terra Group).

EDITOR'S NOTE

Since the writing of this chapter, Atlantic Richfield Corporation has merged with BP Amoco. Prior to the merger, ARCO's interest in Ecuador's Villano Field was acquired by its partner, Agip.

REFERENCES

ARCO. 1990. *Guidelines for Exploration in the Tropical Rain Forests of Eastern Ecuador.* Safety, Health, and Environmental Protection Department, ARCO International Oil and Gas Company, Plano, Texas.

Dita, D., and J. Ranganathan. 1997. *Measuring Up: Toward a Common Framework for Tracking Corporate Environmental Performance*. World Resources Institute, Washington, D.C.

International Petroleum Industry Environmental Conservation Association (IPIECA) and the Oil Industry International Exploration and Production Forum (E & P Forum). 1997. *The Oil Industry: Operating in Sensitive Environments*. IPIECA and E & P Forum, London.

Lindstedt-Siva, J., L. Soileau, D. W. Chamberlain, and M. L. Wouch. 1995. *Engineering for Development in Environmentally Sensitive Areas*. Engineering within Ecological Constraints. National Academy of Engineering.

National Research Council. 1982. *Ecological Aspects of Development in the Humid Tropics*. National Academy Press, Washington, D.C.

Mendez, S., J. A Parnell, S. Reider, and R. F. Wasserstrom. 1998. Model for relations with indigenous people key to ARCO's Ecuadorian rain forest oil project. *Oil and Gas Journal*.

Mendez, S., J. A. Parnell, and R. F. Wasserstrom. 1998a. Seeking common ground: Petroleum and indigenous peoples in Ecuador's Amazon. *Environment* 40(5).

————. 1998b. Technical environmental committee: Finding common ground for E&P operations in indigenous communities. Paper presented at the Society of Petroleum Engineers International Conference on Health, Safety, and Environment in Oil and Gas Exploration and Production, Caracas, Venezuela. June 8–11.

Woodward Clyde International. 1992. *Environmental Impact Evaluation, Proposed Exploratory Well Villano No. 3, Villano, Oriente Basin, Ecuador*. Woodward Clyde International, Denver, CO.

6 Monitoring Impacts of Hydrocarbon Exploration in Sensitive Terrestrial Ecosystems

Perspectives from Block 78, Peru

Jorgen B. Thomsen
Carol Mitchell
Richard Piland
Joseph R. Donnaway

In 1996, Conservation International (CI) and Mobil Oil Corporation entered into a partnership to explore ways of improving "best practices" for energy development in sensitive terrestrial ecosystems. The impetus was, on the one hand, the fact that Mobil was in the process of acquiring an exploratory oil concession in an area where CI had been involved for some time in participatory conservation planning, and on the other hand, the recognition that the system of environmental impact assessments (EIAs) and social impact assessments (SIAs) was not necessarily enough of a safeguard in an area of high biological and social complexity. The system of impact assessment is generally required by law but is often seen as a one-off activity that focuses on construction activities and microlevel alteration of the landscape. CI and Mobil felt that an ongoing and long-term evaluation or monitoring of environmental and social impacts at the ecosystem level was required to ensure that the area's ecological and social integrity was upheld.

The CI-Mobil relationship has since taken the form of a series of structured and unstructured dialogues focusing specifically on oil exploration in the Peruvian Amazon. Toward this end, CI, has designed a project aimed at developing an ecosystem-based ecological and social impact monitoring system for oil exploration activities in what is known as Block 78, a rain forest area of 1.5 million hectares superimposed on the ecosystem of the Tambopata-Candamo Reserved Zone (TCRZ) and sandwiched between the Bolivian border and the border of Peru's Manu National Park. This project has become known as EISA, from its Spanish acronym.

Figure 6.1   Peru.

The mutually agreed goal of EISA is "to develop a long-term, adaptable methodology to measure and evaluate the social and ecological effects of petroleum exploration and production in the Peruvian Amazon." At the same time, information should be "captured and applied to minimize negative impacts and enhance benefits" during the phase of petroleum exploration evaluated. The work is intended to "support a model of the coexistence of petroleum exploration and production with social and ecological health in sensitive ecosystems worldwide"

(CI Peru 1996). While EISA is focusing on both environmental and social monitoring, in this chapter we discuss primarily the former.

These goals are lofty, but they are worthy in their attempt to reconcile the inherent conflicts of interest between petroleum exploration and production and the preservation of the health of Amazonian ecosystems. As part of the development of such a model monitoring system, it is important to understand these conflicts of interest and put them into perspective for Amazonia.

In one often-cited paper on petroleum development in sensitive ecosystems, an attempt was made to uphold a "protection" concept for oil concessions, citing functioning oil fields in California as prime examples of how petroleum development assists in biodiversity conservation. A call was made to apply this concept to rain forest oil exploration and production sites (Lindstedt et al. n.d.). There is definitely a positive aspect for conservation in terms of the restriction of public access to the area of oil development, even though almost inevitably this restriction works better on a local rather than ecosystem scale. However, the direct translation of the "protection" concept to rain forest sites is impossible: In California nearly all natural vegetation was exterminated prior to the advent of oil exploration (certainly prior to the development of the Santa Barbara and Kern County oil fields), but in the Amazonian rain forest the ecosystem as a whole is fairly intact, and much oil exploration is occurring in nearly pristine areas.

Consider the natural environment prior to impact by industrialized society as a delicate silk fabric weaving, a much-used metaphor that in this case is very apt. In the case of California oil fields, there is real protection of the last few threads of the once-unified weaving, which we will never be able to understand fully or reproduce because we never were able to study the pattern. In the case of the Amazonian rain forest, the pattern of the weaving still exists, infinitely more complex than any pattern created by a temperate ecosystem. Colonization, shifting agriculture, and natural resource extraction (timber, minerals) are eating away at the edges of the fabric, and oil exploration (among other extractive activities) is creating holes in the midst of the pattern. Large transnational oil companies are among the few entities with the economic resources to overcome the logistical obstacles of remote, otherwise pristine areas.

This is the context within which a model monitoring system should be developed. For these reasons, it is essential that the oil industry develop protocols, systems, and technologies to reduce and minimize impacts before they occur. Restoration technologies, although of interest, are in a distant second place because knowledge about the ecological processes that create and maintain biological diversity in the rain forest is at a nascent stage.

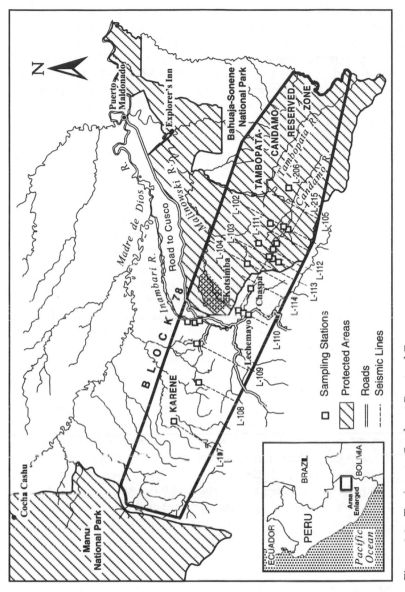

Figure 6.2 Tambopata-Candamo Reserved Zone.

It is clear that some biodiversity loss will always occur when petroleum development projects enter remote ecosystems such as that of the
upper Tambopata. The amount of biodiversity loss will depend much
on (1) the framework of environmental law regulating petroleum exploration and production; (2) the extent to which exploration leads to production; and (3) the environmental attitude of the petroleum company
operator for the area. Our experience in Block 78 in Peru to date indicates that at this point in time, the latter two issues are the most important in determining the fate of an ecologically sensitive area.

The Local Context

Block 78 is composed of two sectors. The Karene sector makes up the
first half of the block, from the Inambari River west to the border of
Manu National Park. The Tambopata sector is formed by a major part
of the TCRZ, overlapping the headwaters of the Tambopata River.
While the Karene area has been studied very little biologically, Tambopata has long been identified as an area of extremely high biodiversity (CDC 1995; Portillo 1914; Raimondi 1864). This reputation is due
mainly to early natural history expeditions and to studies carried out
in the lowland alluvial forests located along the Tambopata River itself, in the northern part of the Reserved Zone. For example, studies
at the Explorer's Inn in the lower Tambopata resulted in world records for butterfly and bird diversity (CDC 1995).

The TCRZ was created in 1990 to conserve and protect this natural
diversity (Ministerial Resolution #032-90-AG/DGFF). CI began work
in the TCRZ in 1990 and was instrumental in the production of the
original proposal presented in 1994 to the government for the zoning
of the protected area into national park, national reserve, and multiple use zones (Supreme Decree 012-96-AG, Lima, Peru).

In 1993, both the Peruvian government and Mobil Oil became interested in hydrocarbons in Madre de Dios, and a two-year agreement
was signed to explore hydrocarbon resources in the watershed of the
Madre de Dios river, using available data and remote sensing techniques. In 1996, a seven-year contract was signed allowing Mobil to
proceed with exploration activities on the ground in Block 78. As a
result, the zoning project presented for Tambopata in 1994 was modified, and Bahuaja-Sonene National Park was declared in a smaller
area, outside the borders of Block 78. Negotiations among CI's Peru
Program, interested members of the Peruvian Congress, and the Peruvian Natural Resources Institute (INRENA) resulted in the inclusion
of a unique clause in the legal decree that created Bahuaja-Sonene
National Park. If any of the lands originally proposed as part of
Bahuaja-Sonene National Park are released by the oil company during
the course of the exploration contract, these lands will automatically

attain the status of national park. Prior to that event, the Tambopata section of Block 78 retains the status of reserved zone.

Up until 1996, the southern section of the TCRZ, primarily composed of montane forest, was little known scientifically because of the difficulty of access to the area. Biological surveys conducted previously in the TCRZ included sites on the lower Tambopata River, and the Cerros del Tavara up to the confluence of the Guacamayo river (CDC 1995; Piland and Varese 1997; RAP 1994). These studies highlighted the rich biological diversity of the northern part of the Reserved Zone as well as the variability of habitats and species composition among sites.

One of the most important tasks in conservation, and the system of triage in ecosystem survival that has inevitably developed as the human frontier expands, is identifying areas with high biological diversity. Just as high and low biodiversity areas are identified within a country or a continent, the distribution of species among available habitats can vary tremendously in a large area such as the TCRZ.

The presence of oil exploration and the possibility of production in the TCRZ demanded the acquisition of information about the presence, distribution, and ecology of species in the montane forest covering Block 78. It was necessary to learn more about the distribution of biodiversity within Block 78 in order to be able to measure changes. Without this basic knowledge it would be impossible to estimate the ecological impact of petroleum exploration.

Design of the Monitoring Scheme and Factors Influencing the Design

The design of a project to monitor environmental impacts of petroleum exploration must take into consideration from the beginning the different impacts inherent in different stages of a petroleum exploration and development project. This design should include at least two parts: (1) a project to measure short- and long-term changes in biodiversity in the area of interest and (2) a "direct impacts" project that has the short-term flexibility to adapt to study of impacts noticed during a particular phase of operations and the longer-term flexibility to adapt to the different phases of operation.

In order to design the most appropriate sampling regime for the detection of environmental impacts in Block 78, we first analyzed the kinds of impacts that might occur over the life span of the exploration project and beyond. We did this initially at a predictive level through analyzing different direct and indirect impact scenarios that might happen in a project of this nature. To achieve this, we brought key people from Mobil's project management and design team together with knowledgeable tropical biologists and anthropologists.[1] The resulting workshop provided a series of recommendations on indicators and sampling regimes that would be appropriate in the context of Block 78.

After initial visits to the seismic lines, we revised our priorities for measuring certain kinds of impacts that have been considered important in other petroleum exploration operations, such as the direct impact of the opening of seismic lines on the flora of the area or the loss of biodiversity due to hunting by seismic workers. For example, when we were designing the EISA project in August and September of 1996, the total area estimated to be deforested during seismic exploration activities was about thirty hectares, or about the size of a small- to medium-scale cattle ranch scattered in openings of one hectare and less throughout well over a million hectares of the Block. The actual area deforested during seismic operations was 56.65 hectares of canopy deforestation (in helipads, resupply camps, and drop zones), and 94 hectares of understory deforestation (along seismic lines) (Walsh Environmental 1995a; Walsh Environmental 1995b; Walsh Peru, S.A 1997). We observed that, in general, fly camps of seismic workers were well supplied with food brought in by helicopter, and hunting of wild animals was almost nonexistent.

Based on our monitoring activities to date, we note that seismic exploration as it is currently being carried out by Mobil in Peru is probably the least environmentally damaging of the various activities leading to and including the production and transport of petroleum from an Amazonian locality. In an area with unknown biodiversity, it seemed correct to invest resources in the study of those impacts important to the long-term biological health of the Tambopata-Candamo-Karene area. We have prioritized the bulk of our monitoring activities accordingly.

In order to ensure that the monitoring system is capable of detecting ecosystem-level impacts, we designed the system with several spatial scales in mind. Table 6.1 exemplifies the monitoring scheme implemented in Block 78 for detection of environmental and social change.

Toward a Definition of Ecosystem Health: Types of Impacts Considered to Be Important in Block 78

In an area as large as Block 78, impacts may occur over a wide range of scales of time and space. First, large-scale effects on the region will probably only be evident over the long term, on the scale of five to ten years or more. Second, although certain exploration and production activities may only have medium or low levels of impact, these may still be significant for biodiversity. We decided upon a strategy for monitoring in which we sampled the biological environment at both

Table 6.1 Structure of Monitoring System Used in Block 78

Scale	Tambopata-Candamo	Inambari	Karene
Landscape level >500,000 ha	Analysis of forest cover change		
Ecosystem level 50–500,000 ha	Baseline information on selected indicator species of flora and fauna	Detailed species inventory	Detailed species inventory
	Distribution patterns of forest communities	Socioeconomic monitoring of communities	Socioeconomic monitoring of communities
	Socioeconomic profile and mapping of colonist and indigenous communities		
	Analysis of land-use patterns		
	Detailed species inventory		
	Analysis of regeneration patterns		
	Detailed description of forest types and vegetation patterns		
	Analysis of landslide frequency		
	Socioeconomic monitoring of communities		
	Water sampling and analysis of contamination		
Species level <50,000 ha	Volumetric analysis of timber species	Volumetric analysis of timber species	
	Analysis of socioeconomic changes in towns	Analysis of socioeconomic changes in towns	

levels. These levels correspond to the two required parts of a monitoring project, as described earlier.

Large-scale impacts. These can be indirectly or directly related to oil exploration activities. Indirect large-scale impacts can result from increased economic activity linked to the presence of the oil industry in the region: for example, migration of people to the area and subsequent deforestation, decline of animal populations, and general declines in biodiversity. These large-scale effects may happen as a result of an increase of *any* economic activity in the region, whether it is oil exploration, mining, agriculture, or other activities. Joint interpretation of our biological data during the next several years with the data obtained by EISA's social team is crucial if we are to be able to interpret the basic reasons for large-scale ecological change. Direct environmental impact such as that produced by a large oil spill and contamination of a major river system is also defined as large-scale impact.

Medium- and small-scale impacts. These occur due to specific activities related to oil exploration, such as opening of the canopy along seismic lines, drop zones, and helipads; use of water sources for supply or disposal of materials (chemicals and other kinds); behavior of oil workers; noise pollution; and so on. In order to investigate these effects, we continually informed ourselves of the current, near future, and future activities of Mobil and its subcontractors during the seismic stage. Based on this information, previous field experience, and observations from the field in Block 78, we defined environmental problems that could result from the seismic stage of exploration in the Tambopata and Karene areas.

Specific Goals for the Seismic Stage of Exploration

With the general goal and the environmental context in mind, we developed the following specific monitoring goals for the baseline phase of EISA:

- Create a baseline data set for long-term environmental monitoring in Tambopata and Karene. This goal was developed in response to the need to know more about the composition and distribution of the flora and fauna in Block 78. We approached this problem through quantitative surveys of workable taxonomic groups: plants, birds, mammals, herpetofauna, and fish.
- Design and carry out studies of the medium- and small-scale impacts observed during the seismic operation. For the seismic stage of exploration in Block 78, we developed three such studies:
  - Erosion potential of seismic lines, drop zones, and helipads;

this includes study of natural regeneration of vegetation in
drop zones and helipads;

- Effect of helicopter noise on "game" animals; and
- Use of seismic lines to access otherwise inaccessible natural
  resources.

- Develop a feedback system for information to Mobil about the
  results and any suggested changes in operations. This goal was
  achieved gradually throughout the project via frequent meetings with Mobil personnel and subcontractors, the production
  of preliminary reports, and the production of a final synthesis
  based on the collected baseline information. The synthesis was
  accompanied by a database containing all biological and social
  baseline data in their raw form.

Baseline Biodiversity Data: Measuring Change over Time

As mentioned, one of EISA's goals was to obtain diversity and density
records for certain taxonomic groups during the initial phase of oil
exploration. In a later phase of oil exploration activities, perhaps a
year or two later, a second measurement of diversity and density will
be required for certain indicator taxa. If changes in diversity or density of organisms are detected then or in subsequent monitoring cycles, we may begin to analyze the possible impacts of oil exploration
on biodiversity. We cannot ignore the possibility that the flora and
fauna may be reacting to other natural or artificial factors. Nevertheless, analysis of the changes may give us a clue to the causal factor.
For example, if there is water pollution, changes may occur in the
composition and abundance of groups such as fish, amphibians, and
certain aquatic insects. If colonization and hunting are the major impacts on the ecosystem, groups such as the large primates and birds
of the family *Cracidae* are likely to show declines in population density or even local extermination.

While complete baseline inventories are ideal, they may take many
years to complete and may not be a realistic objective during the exploratory phase of an oil operation. Taking advantage of the scientific
opportunities provided by seismic lines that are cut through a cross-
section of vegetation types and topographical features (in particular
altitudinal changes), we have used repeated Rapid Assessment Program (RAP) techniques as an alternative to very detailed baseline inventories for fauna and flora. This approach provides a simplified,
up-to-date, scientific appraisal of each site surveyed.[2] When repeated
at distinct sites within the area of interest, the results approximate
the actual diversity of the area.

Table 6.2 Number of Species of Each Taxa Documented at Sites in
Different Localities in Southeastern Peru

Locality	Non-flying Mammals	Birds	Amphibians & Reptiles	Fish	Years of Research
Block 78[1]	46	421	75	101	<1
Lower Tambopata:					
Explorer's Inn[2]	54	438	144	140	>12
Cocha Cashu, Manu[2,3]	70	435	146	—[5]	>17
Pakitza, Manu[4]	62[6]	415	128	210[5]	7

1. This study.
2. RAP 1994, CDC 1995.
3. Gentry 1990.
4. Wilson, D. C. and Sandoval, A., 1996.
5. This figure includes study sites along most of the length of the Manu River and should be considered as the number of species for Manu Park and the Reserved Zone as a whole.
6. Mammals were studied over a four-year period at Pakitza.

Our results for mammal and bird communities in Candamo compare well in terms of diversity to sites such as Cocha Cashu in Manu and Explorer's Inn in Tambopata, which have been under study for more than two decades. Herpetofaunal and fish communities recorded by our team were somewhat less diverse. In the case of herpetofauna, an increase in research effort will probably result in an increase in the number of species found. For fish, the community may be less diverse than fish communities in lower altitude areas (Pakitza, Cocha Cashu, Explorer's Inn) because of the reduction in number of available aquatic habitats (e.g., lack of oxbow lakes).

Direct Impacts: Methods and Preliminary Results

The specific studies of direct impacts of seismic exploration had two sources of origin; either they were considered important before the project started, or they became significant issues as we gained more information during the course of the project. Two issues likely to be important, the regeneration of helipads and the use of seismic lines by people, were predicted before the project started. As we began discovering the ecological characteristics of Block 78, particular issues such as the threat of increased erosion along seismic lines, as well as in helipads, began to come forward in terms of importance for direct impacts. Several months into the project, after field visits by the social team of EISA had been conducted, it became apparent that the issue of helicopter noise was important to the people who lived in the vicinity of seismic work.

Erosion. In the montane forest of the upper Tambopata and Inambari rivers, steep slopes are subject to a high natural rate of landslides.

The most probable environmental effect of the opening of helipads and drop zones is likely to be erosion, in the form of landslides occurring during the rainy season. Landslides may also result from dynamiting on steep slopes if underground movement loosens the overlying soils enough to allow a high rate of water penetration during rains.

Helipads in these steep areas are often put on a hilltop because that location is the safest for landing and takeoff. A denuded hilltop, however, may provide all that is needed for landslides to take place. The Environmental Management Plan for the seismic phase of Block 78 included treatment for helipads to be undertaken after all seismic preparation and data collection crews had completed their work in the area.

Our study of erosion had two components. The first was an aerial video survey of seismic lines around the Candamo Valley.[3] The second was the compilation of information on regeneration of helipads and drop zones based on the monitoring team's direct observations. We obtained aerial video footage of six seismic lines after the rainy season of 1997 (May 1997). Control lines were flown parallel to the seismic lines at a distance of approximately five hundred meters. Original video footage was analyzed, and contiguous images were captured and recorded on compact diskette.

Estimates of new landslide area in square meters or hectares for seismic and control lines is the main output of this analysis. However, the analysis ended up being much more complex than was originally thought. The measurement of the area in the video frame depends on a number of factors such as the distance between the ground and the video camera (e.g., airplane), which in mountainous terrain is constantly varying, and the slope of the particular hill being photographed. Detailed topographic information provided by Mobil is assisting in solving these technical difficulties.

The rate of landslides undoubtedly varies between years, depending on climactic conditions. It is possible that even a short-term increase in landslide rate caused by seismic exploration is not out of the ordinary for the montane ecosystem in Block 78. In order to determine if landslide rates are different from what the ecosystem normally receives, the rate of landslides during the experimental period must be compared with the normal rates of landslides within the area. To ascertain average landslide frequencies, we are studying aerial photos of the Tambopata watershed taken in the 1960s, 1970s, and 1980s in order to estimate "normal" variation in landslide rates with time. Seismic and control lines will be superimposed on these photos, and numbers and area of recent slides will be calculated for those time series.

Regeneration of helipads. Depending on the general topography of the area, helipads were located between three and four kilometers apart along the seismic lines. At each helipad there was a fly camp for workers to stay in during the different phases of seismic exploration. The distance from the helipad to the line varied; some helipads were actually located on the seismic line, and others were located as far as one kilometer away if no suitable place was found closer.

Within each helipad we distinguished three different zones. The central zone (zone 1) is the square area adapted as a direct landing pad, usually not larger than five by five meters. This zone is prepared to be completely flat, with severely compacted soil. On some hilltops, a portion of this area was shored up using cut stems from the slash from the opening of the helipad. Zone 2 is the area around the landing pad, which must be kept fairly free of vegetation for safety reasons. This area may vary in size from about twenty by twenty meters to nearly forty by forty meters. Zone 3 is the border between the helipad and the forest, where trees were cut and the slash from the cleared central area is stored.

We observed a total of twelve helipads in some state of regeneration. In general, the small zone 1 showed little to no regeneration at all: this was true both for the helipads that had been "restored" by Mobil field workers and for those that had experienced no restoration. Zone 2 tended to show the best successional vegetative regeneration, probably because this area was exposed to sun and was not compacted. Regeneration in zone 3 tended to occur mostly through resprouts of cut trunks and the intrusion of vines from surrounding forest. Regeneration of early successional species was scarce in zone 3, probably because of the presence of slash covering the soil. Since the variation in time since abandonment, season, and other factors were high among the helipads visited, a good way to present this variability is by describing several helipads.

Line 102, helipad 11 (November 1996). Line 102 was partially closed in October 1996, but the southern section, where helipad 11 was located, had not been worked over by the restoration crew. Helipad 11 was larger in size than any other helipad visited, probably due to the fact that the fly camp was established at one end of the helipad. Regeneration between October and November was relatively good, especially in zone 2, where abundant seedlings of *Alchornea triplinervia*, and fewer of *Cecropia, Jacaranda*, and one legume species, were found to reach about forty centimeters in height. On the central platform (zone 1), no seedlings were present, and in the area of slash at the helipad border, regeneration was not abundant and generally consisted of resprouts of cut *Heliconia* and palms and the entrance of lianas into the clearing.

Line 103, helipad 11 (December 1996). Located about three kilometers from subbase camp 2, at an altitude of three hundred meters, helipad 11 was closed for a minimum of one month before our visit. The surface is about fifty by eighty meters, larger than most of the other helipads visited. Zone 1 was filled in with white sand brought in from surrounding areas. At the time of the visit in December 1996, this zone was covered with branches and slender trunks, and only three seedlings were observed. In zone 2, regeneration was scarce, reaching a height of only twenty to thirty centimeters. The most abundant seedlings were *Fabaceae* (legume); other species such as *Alchornea triplinervia* and *Manihot brachiloba* were present in lesser quantities. Resprouts of *Bactris, Inga, Heliconia,* and *Olyra* were also present in zone 2. Zone 3 had little regeneration, consisting mostly of tree and palm resprouts.

Drop zones. Small clearings that are located at intervals of approximately five hundred meters along the seismic line, drop zones are used for placement of heavy cargo during the different phases of seismic exploration. The size is usually not more than ten meters on each side, often less. There is less compaction of the soil here than in the central platforms of the helipads. We have evaluated a total of twenty drop zones along three different seismic lines. The drop zones on lines 102 and 103 had been closed by the restoration crew; those along line 110 had only been temporarily abandoned and had not received treatment.

The regeneration among drop zones was quite variable; in fact, each appeared to have its own pattern. In many cases, the dominant regeneration was from cut stems, especially of monocotiledons such as grasses, *Musaceae* (banana-type plants), *Maranthaceae,* and palms. The genus *Inga* (a legume) responded best among the dicotiledonous plants. At least one species of fern also appeared to respond well in terms of regeneration. It is likely that these species were predominant in regeneration because many of the drop zones were probably originally established in gaps. Although this strategy may not have been planned, opening of a small clearing in a gap is easier than in closed canopy forest, and seismic topographic crews probably searched out such sites for drop zones.

The drop zones visited did not show any sign of erosion, and regeneration was good, taking into account the small size of the clearings. The activities of the restoration crew, in terms of covering the drop zones with slash, do not appear to have any beneficial effect over these areas. The small size of these clearings, which approximates the size of clearings produced by small natural treefalls and branchfalls in the forest, would suggest that they should regenerate quickly.

Seismic lines. Between October 1996 and January 1997, a total of ten kilometers of recently closed seismic lines (lines 102 and 103), more than twenty kilometers of recently abandoned lines (line 110), and more than thirty kilometers of lines in activity (various lines) were observed.

In very few cases were the Environmental Management Plan guidelines not followed in the creation of seismic lines: trails were generally 1.5 meters wide, and trees more than twenty centimeters in diameter were rarely cut. When this did happen, it appeared to depend almost entirely on the criteria of individual topographers. During our study visits to topography campsites, it appeared that some topographers made distinct efforts to reduce the cutting of vegetation, whereas others were not so concerned. In tall forest, the overhead canopy was not broken, and the line was only visible from the air in those habitats where the vegetation is naturally low or broken (e.g., dwarf forest, very steep slopes with ferns, palm swamps). A positive observation was that the stakes (i.e., cut tree stems) used to mark topographic sites along the seismic lines in some cases were beginning to sprout leaves and were likely to be rooting.

Other phases of seismic exploration appeared to result in some negative effects. Pieces of plastic cable and other plastic trash were not uncommon along trails closed by the restoration crew. However, we consider the most potentially damaging effects to be the explosive "blowouts," holes in the forest floor caused when a drill hole is inadequately tamped and blows "out" when the explosion takes place. Our botanical team observed several of these, the largest of which was about a meter in diameter and eighty centimeters deep. We consider it possible that blowouts located on steep slopes could contribute to higher erosion or landslide rates along seismic lines. This potential effect might be detected when a second videography study is conducted.

Helicopter noise. During seismic operations, small to medium-sized helicopters (those that carry from four passengers to those that carry about ten plus cargo) are used to transport equipment and supplies from the base camp into the field. This mode of operation, often called the offshore model, is an alternative to the creation of a road network into the area of interest. Local people in the rain forest, including native people and colonists, insist that the noise of the helicopters drives away game animals from their hunting areas. Even before seismic operations began in Block 78, the problem of noise pollution was discussed among conservationists, ecologists, and Mobil personnel (CI Peru 1996). The information presented here does not resolve the problem of noise pollution but rather brings forward certain issues that need to be taken into account for a thorough evaluation of the problem and preparation of management plans.

In the first months of the environmental work in Block 78, the issue of noise pollution was not addressed for two reasons. First, most of the seismic work was taking place in the unpopulated areas of the block: the eastern part of the Candamo Valley and Távara Hills. Not until December 1996, when line 110 was opened in the Lechemayo/Chaspa area, were helicopters flying over areas where people hunt regularly. In the first months of 1997, line 110 was planned to pass through the native community of Kotsimba.

Second, to the biologists with EISA's ecological team, it did not seem logical that all game animals should be frightened away by helicopter noise. Particular species perhaps, such as the giant otter (*Pteronura brasiliensis*), tamarin (*Saguinus* spp.), squirrel monkey (*Saimiri sciureus*), and some bird species, may be predisposed to disturbance by helicopter noise because they normally communicate vocally using high-frequency sounds. However, these species are not preferred as game species, which are represented by the tapir (*Tapirus terristris*), peccaries (*Tayassu* spp.), deer (*Mazama* spp.), large primates—spider monkey (*Ateles paniscus*), woolly monkey (*Lagothrix lagotricha*), and howler monkey (*Alouatta seniculus*)— and birds of the family *Cracidae*. Animals are known to habituate to noise of almost all kinds, as is shown by the presence of flocks of birds on airport runways and deer feeding next to busy highways in the United States. In Madre de Dios, during the construction of two tourist lodges and one biological station, EISA's ecological team leader observed that fauna of all kinds habituated to the noise of chainsaws, circular saws, and generators in the rain forest. Indeed, one species of bird (*Cacicus cela*) that uses high-frequency sounds to communicate was overheard imitating the sound of a chainsaw as part of its mating song.

Nevertheless, helicopter noise is loud, and it contains sounds covering a wide range of frequencies (Manci et al. 1988). During the field work of EISA's social team, it became clear that helicopter noise was still an issue. One local person, who was living on the Puerto Maldonado-Cusco road twenty-five kilometers north of Block 78, perceived an increase in the number of animals near his garden: his explanation was that the seismic work in Block 78 was scaring all the animals out north toward the road. Moreover, interviews by EISA's social team in Kotsimba revealed that residents were worried about the effect of helicopter noise on fauna. Although it was apparent that helicopter noise had people worried, we decided not to accept these interview data as utter truth without further information, as people's concerns are not always expressed in logical fashion. We decided that it was important to get information about how helicopter noise affects people as well as fauna.

The design of a study of effects of helicopter noise on rain forest fauna is not simple, and a "thought experiment" about how one would ideally like to study the situation is useful to determine which variables are important and feasible to study. One would wish for situations of the following kind.

One ideal study situation would involve two areas, initially equally abundant in fauna, one with helicopter traffic and one without. One could conduct censuses in both areas and, given enough time, could compare the data to see whether the area with helicopter traffic had a reduced density of fauna. Normally, detection of animals in the rain forest is initially made by an auditory (animal calls, movement of branches, pieces of fruit dropping to the ground) and then visual search. Because of this, the effect of helicopter noise on the observer's ability to detect animals would have to be taken into account and estimated. (Note that helicopter noise may also affect a hunter's ability to hear animals and perhaps result in a real reduction of detection of game without necessarily having any effect on the animals at all).

A second ideal situation, simply to measure the reactions of fauna over time, would be an open area with high visibility and high, constant animal visitation rates, where an observer could record the reactions of a number of animals over time to the presence of helicopters. This situation occurs fairly often in temperate or savanna areas, but open areas with high visibility are scarce in the rain forest. The one possibility which we considered and pursued was the use of macaw and mammal clay licks (ccollpas) as points where animals congregate and are visible and at which the reactions of fauna could be observed. One advantage of ccollpas is that they do attract game animals and therefore would be a fairly direct measure of the problem expressed by local people. A second advantage is that all data can be acquired visually, and effects of helicopter noise on observer ability to detect fauna would be minimal.

Difficulties with using ccollpas as observation points in a study such as this are as follows. First, visitation rates of animals to ccollpas are not constant, and undergo seasonal variation (especially bird ccollpas; perhaps less so for mammal ccollpas). Second, the ccollpa would have to be in line with a frequent helicopter flight path in order to qualify as affected by noise. This would happen only by chance, and in a large area such as Block 78, we have the "needle in a haystack" probability of it occurring. Third, ccollpas, especially large ones, are inevitably used by hunters from nearby communities. In supposedly affected areas (i.e., hunting grounds of local communities), the population of fauna arriving at a ccollpa is likely to be severely reduced from its original density, and very wary.

Interviews with six hunters in the native community of Kotsimba, located on the northern border of Block 78, provided information about the prior and actual distribution of particular hunted animal species. Those species described as most affected by helicopter noise were precisely those species also described as most overhunted in the last fifteen to twenty years. All hunters agreed that animals were less abundant now than they used to be. The place described as best for hunting of forest game species was the "cumbre," or the ridge of hills to the south of the Kotsimba community. When asked where the animals had fled to because of helicopter noise, most hunters indicated that they had fled to the "cumbre." It is therefore almost impossible to tease apart the effects of helicopters and hunting in this particular community.

Only one instance of direct observation of fauna when a helicopter arrived and took off from a fly camp was acquired. This consisted of a group of titi monkeys (*Callicebus moloch*) that was feeding on new leaves at the edge of a helipad (helipad 13, line 201) when the helicopter arrived. The observer indicated that the monkeys showed no reaction to the helicopter whatsoever—they continued feeding normally.

Pertinent to this discussion is the interview with the family of miners living on the northern section of line 108. The head of household in this family also mentioned that helicopters were causing animals to flee and said (we quote): "Even my children are bothered by the noise."

Our general conclusion about helicopter noise is that it bothers *people* and that therefore people may blame difficulties with hunting on this irritating addition to their environment. We cannot conclude that animals are or are not directly affected based on the information we have up to this point. However, the fact that people are bothered by the noise should be taken into account by an operating oil company. We believe this monitoring exercise was useful in distinguishing that people may in fact be more affected by helicopter noise than are animals; in this case, work to reduce the problem should focus on people rather than fauna.

Human Use of Natural Resources Accessed by Seismic Lines

Most human use of forest resources occurs along well-defined access routes such as rivers, streams, or trails. Seismic lines constitute new, well-made, and straight trails through the forest for many kilometers. We considered the possibility that local people living near seismic lines would gain access to new areas for resource extraction. The principal resource likely to be extracted is timber, which constitutes an important source of cash for many local people in the Block 78 area. Furthermore, at least one of the seismic lines created by an oil company during exploration in the Tambopata area in the mid-1980s is still in use by local people as a route between the upper Malinowski River drainage system and the road to Puerto Maldonado. Students in the community of Kotsimba reportedly also used an old seismic line to reach their school.

Our main approach to monitoring people's use of seismic lines was the initiation of a forestry inventory along line 110 in the Lechemayo/Chaspa area. In addition, EISA's social team conducted interviews with people in the community of Kotsimba, in the small communities where we worked along the Inambari River, and with one family living near our study site on line 108.

Results of interviews indicated that initially people thought that the lines would not be of much use because they follow a straight compass route independent of rough topography, whereas local trails tend to follow the contour lines of hills, seeking out an easy route. Few places in Block 78 have the flat topography that would make use of the seismic lines easy. However, in early 1997, a Mobil environmental health and safety advisor photographed a site near line 110 where a mahogany tree had been felled after the creation of the line. Nonetheless, as one would expect, in any particular instance it is difficult to determine whether a given tree would have been felled had the line not been created.

The family living on line 108 provided the best short-term information on how people may use and take advantage of the leftovers of seismic exploration. The northern part of line 108 is in terrain alternating between flat alluvial plain, near the headwaters of the Colorado River, and steep hills and rock walls. Our field visit occurred after all seismic crews had been through the area, and the line had recently been abandoned. In our conversations with the head of household, he indicated that he intended to use the seismic line to the south to search out new areas for hunting and to look for new sites along streams appropriate for gold mining. The southern section of this line was hilly, and access to this area previously had been limited to walking along stream edges. This family had already taken advantage of the helipad opened close to their home, removing the slash placed on the helipad and readying the area for planting of annual crops. In their opinion, Mobil's restoration crew had "ruined" the helipad by covering it with slash.

Forestry inventory. The main goal of the inventory of timber resources was to determine whether or not people living in the area of Block 78 use the seismic lines to gain access to previously unused areas for resource extraction. After initial visits to the upper Inambari River valley and observation of seismic lines and local economic activities, we began a detailed study of timber extraction in the area as part of monitoring long-term impacts of oil exploration. Three forestry inventories were conducted to measure timber resource potential along seismic lines.

Four forestry concessions of one thousand hectares each were located in the study area: three belonged to local communities, and one was being utilized by an unknown harvester.[4] One timber species, *Cedrelinga catenaeformis*, represents the most important economic species for loggers in the study area. This species is generally recognized as valuable and brings a slightly higher price than other species.

Plots were established along seismic lines 110 and 201, in the vicinity of the Chaspa River. Plots included all forest types that had timber species with diameters greater than thirty-five centimeters dbh. This diameter was taken as the absolute minimum size for eco-nomic production within a forest concession. We worked whenever possible with at least one member of the local community in order to obtain information about common names and identification of spe-cies actually extracted (as compared to species potentially valuable for extraction). We also obtained from these local lumbermen infor-mation about the number of usable boles (volume) from each poten-tially valuable timber tree.

Several conclusions can be made about the logging activity in the upper Inambari area. First, the area is ecologically unsuitable for tim-ber extraction. In spite of reasonably high measures for potential tim-ber volume, most of the timber is in smaller size class trees that may not yet be reproducing, and most of the timber is located on steep slopes where erosion is a distinct risk, considering the extraction technologies used. Second, the extraction techniques used and the severity of the terrain mean that only about one-fifth of the available wood volume is actually harvested. Difficulties in transport of the wood from the harvest site to the point of sale also involve loss of timber (people living along the Chaspa River body-surf their cut beams downstream when the river is up, an incredibly risky process).

However, timber extraction is one of the very few activities that produces cash for local people, and it is unlikely that extraction will cease before all available timber is cut. In and around line 110, and perhaps in other areas too, the increased access provided by aban-doned seismic lines could result in an increase in timber extraction. Because of the present reproductive characteristics of the forest, this extraction is unlikely to be sustainable. Therefore, future evaluation of timber extraction and the use to which seismic line 110 may have been put is a necessary component of long term monitoring of im-pacts in Block 78.

Discussion

The first phase of the monitoring project in Block 78 has focused on the area as a whole, with baseline data being sampled on lines and helipads that were opened up for the purposes of seismic data collec-tion. The value of the biological information obtained goes far beyond the specific purposes for which it was collected. The scientific under-standing of the rain forest contained within Block 78 has been greatly increased. Many new records of biota previously unknown from this area or unknown to science have been sampled and attest to the global significance of the area from a biodiversity conservation per-spective.

Some of the key observations that we see emerging from this project underscore the fact that in sensitive and poorly explored terrestrial ecosystems, one cannot rely only on EIAs, as they are traditionally carried out, to provide the biological and ecological information required for ensuring that the health of the ecosystem is maintained. A greater up-front investment is required to ensure that enough information is available for project management purposes. This, in turn, means that managers of oil and gas exploration projects must factor the need for ongoing data gathering into project time frames and allow enough flexibility in the project execution phase to take full advantage of such data.

From our perspective, the following lessons learned have broad application.

- In order to fully understand operational practices employed by the oil company and design an ecological monitoring system that will accurately evaluate operational impacts, it is important to be involved in the operational planning phase from the beginning and to have access to operational management plans.
- A detailed assessment of impacts that are already present (e.g., small-scale gold mining activities that use mercury) prior to the arrival of the company whose activities are being monitored is essential in order to distinguish impacts and attribute them to the correct source.
- In setting indicators for monitoring impacts of resource extraction, a careful predictive analysis (using a logical framework approach) of probable impacts is required in order to select and focus field methodologies and sampling techniques. In other words, it is important to have a fairly good idea of the probable impacts the monitoring scheme is designed to detect and over what time frame they are likely to occur.

As Mobil's involvement in Block 78 moves into the exploratory drilling phase, the monitoring project will go into its second phase, with additional data collection to cover the possible impacts of drilling. The information we obtained during the seismic phase is of great use in the development of appropriate studies during the drilling phase. In addition to rapid assessment techniques, methodologies that are more geographically focused will be employed (CI 1997b).

Acknowledgments We thank the Mobil Foundation for financial support, the EISA ecological team members for their excellent field work, and the CI-Peru Program staff for help in every way. The work was done under authorizations from INRENA (#86-96-INRENA-DGANPES-DANP; #24-97-INRENA-DGANPES-DANP; #570-96-PE).

EISA is made possible through a grant from the Mobil Foundation. Permits for our work within the Tambopata Candamo Reserved Zone were obtained from the National Institute for Natural Resources (INRENA) and the Ministry of Fisheries in Lima (#86-96-INRENA-DGANPES-DANP; #24-97-INRENA-DGANPES-DANP; #570-96-PE).

We are grateful to all the many people who have contributed to the project. In particular, we would like to acknowledge the contribution of EISA's field staff: Fonchii Chang (ichthyofauna), Fernando Cornejo (plants), Jorge Luis Hurtado (herpetofauna), Javier Icochea (herpeto-fauna), Lawrence Lopez (birds), Talia Llosa (giant river otters), Eddy Mendoza (cartography), David Naish (large mammals), Gabriela Ori-huela (large mammals), Juan Pesha (large mammals), Tom Schulen-berg (birds), Juana Silva (cartography), José Tello (birds), and Horacio Zeballos (small mammals).

EDITOR'S NOTE

Since the writing of this chapter, Mobil Corporation has merged with Exxon Corporation to form Exxon Mobil Corporation.

NOTES

1. CI workshop report.
2. For a description of rapid assessment methodologies, see Piland and Varese (1997).
3. For methodology, see Conservation International (1997). Conservation International. 1997. *Evaluation of Social and Environmental Impacts of Petroleum Exploration in Block 78-Tambopata. Baseline Biodiversity Data and Studies of Direct Environmental Impacts of Seismic Exploration.* Final report. Conservation International, Washington, D.C., and Lima, Peru.
4. For a detailed description of the forestry concessions, see Manci et al. (1988).

REFERENCES

CDC. 1995. *Reporte Tambopata: Abstracts of Investigations around Explorer's Inn.* Centro de Datos para la Conservacion, Universidad Nacional Agraria de La Molina, Lima, Peru.

Conservation International. 1997. *Ecological Characterization of the Candamo Valley, Peru — Identification of Ecosystem-wide Information Relevant to Impact Assessment of Exploratory Drilling for Hydrocarbons.* Conservation International, Washington, D.C., and Lima, Peru.

Conservation International—Peru. 1996. *Memoria del Taller sobre Evaluacion de Impactos Sociales y Ambientales de la Exploracion para Hidrocarburos en Tambopata.* Atinchik, Lima, Peru.

Gentry, A., ed. 1990. *Four Neotropical Rainforests.* New Haven, Yale University Press.

Lindstedt-Silva, J., L. C. Soileau, D. W. Chamberlain, M. L. Wouch. No date. Oil operations in a rain forest. Reprint from *Engineering within Ecological Constraints.* National Academy of Engineering, Washington, D.C.

Manci, K. M., D. N. Gladwin, R. Villella, and M. G. Cavendish. 1988. *Effects of Aircraft Noise and Sonic Booms on Domestic Animals and Wildlife: A Literature Synthesis.* U.S. Fish and Wildlife Survey, NERC-88/29. National Ecology Research Center, Ft. Collins, Colo.

Ministerial Resolution no. 032-90-AG/DGFF; 148-92-AG, Lima, Peru.

Piland, R., and M. Varese. 1997. *Memoria del Programa de Desarrollo basado en la Conservacion en Tambopata (PRODESCOT).* Conservación Internacional, Programa Peru, Lima, Peru.

Portillo, Pedro. 1914. Departmento de Madre de Dios. In Sociedad Geográfica de Lima. Vol. 30, books 1 and 2.

Raimondi, Antonio. 1864. Expedicíon a las quebradas de Sandia y Tambopata. In *El Perú*. Vol. 1, pt. Imprenta del Estado, Lima, Peru. Chapter 5.

RAP. 1994. *The Tambopata-Candamo Reserved Zone of Southeastern Peru: A Biological Assessment*. Rapid Assessment Program Working Paper no. 6, Conservation International, Washington, D.C.

Walsh Environmental. 1995a. *Apéndice al Estudio de Impacto Ambiental de la Campaña Sísmica en el área de Tambopata, Lote 78*. Prepared for Mobil Exploration and Producing Peru, Lima, Peru.

———. 1995b. *Environmental Impact Study: Karene Seismic Project*. Prepared for Mobil Exploration and Producing Peru, Lima, Peru.

Walsh Peru, S.A. 1997. *Reporte Final sobre la Campaña Sísmica de Mobil Exploration and Producing, Peru en el Lote 78, Tambopata*. Lima, Peru.

Wilson, D. C., and A. Sandova, eds. 1996. *Manu: The Biodiversity of Southeast Peru*. Smithsonian Institution, Washington, D.C.

PART III

Forests under Pressure

7 Trade, Transnationals, and Tropical Deforestation

Nigel Sizer

From Guyana to Gabon, from Suriname to the Solomon Islands, grow-ing global demand for wood and its derivatives has been driving an onslaught on the world's remaining tropical forest frontiers. These forests are home to the majority of global nonaquatic biological diver-sity. Between half and three-quarters of global terrestrial biological variety is found living in and depending upon tropical forests (Reid and Miller 1989; Bryant et al. 1997).

Forests also store vast quantities of carbon. Fully 15 to 20 percent of global greenhouse gas emissions arise from tropical deforestation and the associated burning and accelerated decomposition. As the world starts to talk about spending money to reduce carbon emis-sions, it is apparent that the cost of such burning adds up to many billions of dollars each year (Brown P. 1998).

Furthermore, a vast diversity of human culture thrives in forests. In each of the Amazon and the Congo river basins, three or four hundred distinct languages and cultures have been described. These are being lost as fast as the forest itself (Lynch 1992).

According to the United Nations, forests cover 3.4 billion hectares (8.5 billion acres) of the earth, or 27 percent of land surface. Between 1980 and 1995, nearly 200 million hectares of forest was lost, and partially offset by reforestation, resulting in a net loss of about 180 million hectares, or an average annual loss of 12 million hectares (FAO 1997). The equivalent of about 3 percent of global forest area is in plantations, an increase of 40 percent in fifteen years, and a 300 percent rise in the tropics. Fire-related deforestation is also on the rise, with massive El Niño–related burnings in the past year in Indo-nesia, Brazil, and elsewhere in the tropics.

These statistics understate the real extent of decline of primary forests. Almost half of the world's original forest cover is gone, much destroyed in the past thirty years. Furthermore, just one-fifth of the world's original forest cover remains in large tracts of relatively pristine *frontier forest*. Seventy-six countries have lost all of their original frontier forest, and three countries—Russia, Canada, and Brazil—house 70 percent of the remaining frontier forest (Bryant et al. 1997).

Outside of boreal regions, 75 percent of remaining frontier forest is threatened, and commercial logging poses a direct threat to these areas in all major regions—Amazonia, the Congo Basin, Indonesia, Papua New Guinea, and beyond (Bryant, D. et al. 1997). Tropical industrial logging is, in most cases, highly unsustainable. Furthermore, the roads that the loggers build to gain access to timber serve as the entry point for farmers, fuelwood collectors, hunters, miners, local artisianal loggers, and others who add to forest degradation and outright clearance.

A key force driving human hunger for logs from frontier forests is, of course, developed countries' growing economies and consumption of wood and wood fiber. Per capita pulp and paper consumption in developed countries is, on average, ten times that in developing countries, and it is demand for the timbers of the highest values that pays loggers to drive their trucks and bulldozers further into the forest looking for more of the most precious species.

Other key root causes of forest loss include growing populations and demand for land as forests are cleared, especially where soils are more fertile, to make way for food production. Bad economic policies that ignore the costs of soil erosion or watershed destruction also contribute, with logging or clearance of forest often superficially appearing to be the most economically attractive option for development. In fact, if the real costs are factored in, conservation and nontimber uses such as ecotourism may be more profitable. Political consideration and even corruption also play a role, as political leaders make short-sighted decisions that pander to their supporters or generate contributions to political campaigns. This can also lead to illegal logging, as seen in Burma, where the government and rebel groups have both financed a decades-long civil war with illegal logging proceeds (Brunner et al. 1998).

This chapter provides a brief review of trends in wood consumption and trade and an analysis of the factors leading to a growth in transnational logging, led by new companies from southeast Asia. Case studies from four key countries rich in tropical forest are provided, followed by concluding remarks.

Wood Consumption and International Trade

Developed countries are responsible for about three-quarters of global wood use, excluding fuelwood (FAO 1997; WRI 1998). Growth in consumption is, however, fastest in developing countries. African consumption has doubled to 4 percent of total global wood use; South American consumption almost doubled to 7 percent; and Asia transformed the balance, with their use growing from 15 to 21 percent. Developing countries' use of wood-based panels and paper and paperboard, tripled. This profile of current wood use and projected demand raises two major issues for those concerned with the deteriorating state of the world's forests. First, demand to liquidate forests to supply wood will continue to grow. Second, the greatest growth in demand will be in developing countries, where meeting basic human needs in the short term is the priority and very limited resources are therefore available to address the problem.

Only 6 to 8 percent of world roundwood production[1] enters global trade, but this small fraction was worth $114 billion in 1994, up by 75 percent in real terms from 1970. Between 1970 and 1994, exports of industrial roundwood increased 21 percent to 113 million cubic meters, sawnwood exports doubled to 108 million cubic meters, pulp exports doubled to thirty-two million tonnes, and exports of panels increased fourfold to thirty-eight million cubic meters, while paper and paperboard exports tripled to seventy-three million tonnes. Unlike those of almost all other commodities, the prices of basic wood products over recent decades has been increasing, indicating that there is growing real scarcity of the resource. Globalization, higher commodity prices, declining primary forest, and increasing international demand will put great pressure on the remaining frontier forests.

The United Nations predicts that roundwood demand will grow at 1.74 percent per year (FAO 1997). Fiber from intensively managed plantations will become increasingly important and will eventually grow to fill the gap as supplies from natural forests decline and increase in cost. There will be a growing price disparity between products extracted from natural forests and those from plantations, stimulating investment in plantations, especially in the southern hemisphere, where land is cheap and growing conditions are ideal. Furthermore, technology will increase substitutability, eliminating the distinction between pulpwood and sawlogs.

Transnational Companies Logging the Tropical Forest Frontier

Transnational companies are corporations involved in business investments outside their country of origin (von Kirchbach 1983). These companies tend to be larger than local businesses. Consequently, they have greater opportunities for vertical integration, global linkages, and even an influence on international investment patterns. Their resources and expertise often far outweigh those of the governments of the countries in which they are seeking contracts or investments.

Multinational logging companies vary greatly in structure and size. Those based in Europe have been active for several decades, mostly in Africa. These include Groupe Rougier in France, as well as Danzer and Feldmeyer in Germany. A number of younger, but now much larger, multinationals have emerged recently in the dynamic economies of Southeast Asia, especially in Malaysia, Indonesia, and South Korea. These include Rimbunan Hijau, Berjaya, and Samling in Malaysia and Barito Pacific Lumber in Indonesia.

The decision by a corporation to invest overseas is influenced by a complex range of factors, including market trends, government policies, local versus offshore production costs, taxes, raw material supply, economic growth levels, and the size, profitability, and level of capitalization of the company (Khan 1986).

Recently, there has been a shift in the sources of investment in transnational logging operations toward Asia. A number of key reasons explain the overseas expansion by Asian logging companies; these are described hereafter.

Scarcity. Log export bans or tariffs in effect in Malaysia and Indonesia, contemplated for the Solomon Islands, and being introduced in Papua New Guinea are resulting in log scarcity—peninsular Malaysia banned exports of ten species of logs in 1972. Indonesia outlawed all log exports in 1985. The Malaysian state of Sabah temporarily banned log exports in 1993 and planned to permanently ban them by 1996 (Crossley 1996). The Malaysian state of Sarawak has limited log exports and log production since 1993, resulting in a 35 percent decrease in exported logs by volume (Schwartz and Friedland 1992). Thailand, Laos, and Vietnam have also restricted or banned logging and log exports. Companies that process foreign timber thus must look outside Southeast Asia for logs. This group includes not only the Japanese, Taiwanese, and Korean processors, which have historically been the main importers of tropical logs, but also the huge Malaysian and Indonesian timber companies that have diversified into timber processing and now need larger supplies of logs to expand and compete.

Between 1991 and 1995, output of tropical logs by Asia-Pacific International Tropical Timber Organization (ITTO) producer member countries declined by 10 percent, with exports nearly halving over the same period. Many southeast Asian countries now import tropical logs. Malaysia, for example, is now not only the ITTO's largest producer and exporter of tropical logs but by 1995 had risen to be the ninth ranked log importer.

Incentives in home countries for investment in timber processing. Malaysia and Indonesia have used tax incentives and grants to encourage domestic investment in sawmills and plywood, molding, and furniture factories. Rimbunan Hijau of Sarawak, for example, has made plywood processing the core of its corporate strategy and is now Malaysia's largest plywood and veneer producer (Pura 1994a).

More effective law enforcement and tax increases in Malaysia and Indonesia. Current revenue collection in Indonesia totals only about 20 to 50 percent of the potential maximum. While companies can still earn outsized profits from logging in Indonesia, the profit margin has shrunk dramatically over the past decade. The Indonesian reforestation fee (the largest share of forest levies), first imposed in 1980 at US$4 per cubic meter, had risen to about $22 per cubic meter by 1997 (Ministry of Forestry, Jakarta 1994).

In addition, governments in Southeast Asia are now holding logging companies accountable for their illegal logging and accounting practices.[2] In late 1992, fines and jail terms were increased for illegal loggers in Malaysia, and companies were warned that if they are convicted of logging illegally, their equipment will be confiscated and they will be made to pay compensation for felled logs (Hamid 1993). In Indonesia, the forestry ministry has revoked concession permits in the wake of unsound logging practices. The accounting practices of these companies are also being more closely examined.[3] Recently, the Malaysian government has taken steps to recapture the revenues lost due to transfer pricing (Pura 1994b).

New concessions as equity for leveraging loans to finance aggressive takeover bids. A company that secures a promising new timber concession is likely to see its credit rating rise; a new concession can serve as equity for securing multimillion dollar loans.

Fear of losing market share. As international prices for raw logs rise, suppliers that can provide lower priced raw materials for the large milling operations in Asia and elsewhere can increase, or at least maintain, their market share. Suppliers, sensing the rising prices and declining availability, may now be prepared to take relatively greater risks with investments, look farther afield, and pay higher transport costs to protect foreign markets that have been developed through export activity.

International trade accords. Regional and international trade accords may attract companies from elsewhere seeking preferential market access to avert threats to exports from other regions from tariff protection or competition. Central and South America and the Caribbean encompass several such agreements, including CARICOM (the regional free-trade zone in the Caribbean Community) and Mercosul (the Southern Cone countries). The Lomé Convention provides for removal of tariffs (around 10 percent for processed timber products) on most European Union imports from the developing countries covered by the agreement.

Japan's Role in Transnational Logging Operations

Japan is the world's largest importer of tropical logs, although its imports have dropped sharply in recent years from 10.4 million cubic meters in 1991 to 6.6 million cubic meters in 1995 (ITTO 1996). There has also been a subsequent increase in the import of manufactured products from those countries that previously supplied only logs, such as Indonesia and Malaysia. Japan is Asia-Pacific's leading producer and consumer of sawn timber, wood pulp and paper, and paperboard and runs second to China and Taiwan (combined) for consumption of industrial roundwood, veneer sheets, and particle and fibre board (FAO 1995). Japan runs second to Indonesia in regional plywood production but is the region's largest consumer of this product. The country is also the world's largest importer of woodchips (Sar and Sledge 1994). Japan sources most of its logs from the Asia-Pacific region but has recently increased imports from African and South American countries including the Congo, Gabon, Ghana, the Ivory Coast, Brazil, and Honduras. Almost three-quarters of Japan's wood needs are met by imports (Callister 1996). The Japanese industrial sector is dominated by large trading conglomerates known as *sogo shosha*. These trading houses are the world's largest players in the international timber trade. Between 1990 and 1993, Japan imported an average of US$55 billion and exported US$7 billion of forest products per year.

Japanese companies have been implicated in transfer-pricing and other tax avoidance practices in Papua New Guinea. Other problems attributed to companies with Japanese interests operating in Papua New Guinea include failure to complete EIAs, ignorance of environmental clauses in permits, cheating landowners of profits, failure to meet promises made to landowners, raising conflicts between rival landowner groups, destruction of villages, and failure to employ Papua New Guinea nationals (Filer 1997). Japan also appears to have been a destination for smuggled timber sourced from the Philippines and possibly Indonesia (Nectoux and Kuroda 1989; Callister 1992).

Malaysian Investments in Transnational Logging Operations

During the period 1990–93, Malaysia exported US$14.2 billion in forest products, accounting for 23.5 percent of the value of all forest product exports from the Asia-Pacific region over these four years and placing it second in rank behind Indonesia. Despite reductions in log exports in recent years, Malaysia is still easily the world's largest exporter of tropical logs. The Asia-Pacific region is the main market for

Malaysia's wood product exports (FAO 1995; ITTO 1996; Callister 1996).

The majority of Malaysian timber production is now in the state of Sarawak, home to Malaysia's five largest logging companies: KTS, Rimbunan Hijau, Samling Group, STIDC, and WTK. With dwindling local timber reserves, many of Sarawak's privately owned timber giants have expanded into new activities and begun to search for new logging concessions abroad. The first truly transnational Malaysian corporations emerged in the mid-1970s as a result of government takeovers of foreign companies operating in Malaysia. In 1982, Malaysian companies were among the largest foreign investors in Indonesian logging concessions, with sixteen enterprises holding concessions covering 1.8 million hectares, representing an investment of US$41.2 million (Walker and Hoesada 1986). Investment has since spread beyond the Asia-Pacific region.

Malaysian logging companies operating overseas have been involved in a number of controversies. In the Solomon Islands, the license of Malaysian company Kumpulan Emas was suspended following consistent breaches of stipulated conditions (Callick 1995). An employee of the Malaysian company Berjaya allegedly offered a bribe to a minister of the Solomon Islands government (Kürschner-Pelkmann 1995). There have also been reports that seven government ministers were paid bribes by Malaysian logging firms (Gray 1996). Still in the Solomon Islands, there have been reports of conflicts between Malaysian company Maving Brothers and local landowners (World Rainforest Movement 1995).

Indonesian Investments in Transnational Logging Operations

Between 1990 and 1993 inclusive, Indonesia exported 34.4 million cubic meters of plywood alone, worth US$13.4 billion. Exports of industrial roundwood and sawn timber have declined since the early 1980s, reflecting government policies aimed at restricting the export of these products. The total annual cut is set to rise from 36 million cubic meters in 1994–95 to 40 million cubic meters by 1998–99. The national target for plantation establishment during 1983–99 was 4.4 million hectares (Ministry of Forestry, Jakarta 1993).

Indonesian companies are not as active overseas as their Malaysian neighbors. This can be attributed to two major factors. First, Indonesia has far more forest of its own to exploit, so Indonesian firms can grow very large without looking overseas for resources. Second, according to local business executives, the government of Indonesia has actively discouraged them from investing overseas, choosing instead to promote domestic investment to stimulate local growth and employment.

Most Indonesian foreign investments, with the exception of two recent ones in Suriname, are in the Asia-Pacific region.

South Korean Investments in Transnational Logging Operations

South Korea imports most of its timber. Around half of its imports of industrial roundwood came from within the region, but only around a quarter of its veneer, particle and fibre board, and paper and paperboard and very little of its wood pulp imports. South Korea was the third largest importer of tropical logs among ITTO member countries from 1991 to 1995 (ITTO 1996). Imports over this period declined from 3.7 million cubic meters in 1991 to only 1.9 million cubic meters in 1995. While most tropical logs were sourced from Malaysia and Papua New Guinea, non-Asia-Pacific countries such as the Ivory Coast, Gabon, Ghana, and Guyana have also exported logs to South Korea in recent years.

There have been various claims of negative impacts of Korean logging companies in the countries where they operate. In the Solomon Islands, Hyundai was ordered in 1993 to pay compensation to local communities on the island of Vella la Vella for illegally taken logs and for environmental damage (Bohan et al. 1996; Newell and Wilson 1996).

Hong Kong and Transnational Logging Operations

Hong Kong is an importer of forest products and serves as transit point for timber imports into China. Paper and paperboard, and plywood, were the two highest-value wood product imports in 1990–93 inclusive (US$5.8 billion and US$1.3 billion, respectively). The highest-value exports over the same period were paper and paperboard (US$2.5 billion), plywood (US$712.4 million), and sawn timber (US$256.9 million). There was strong growth in the import and export of most timber products over the decade to 1993 (Callister 1996). During 1992 and 1993, nearly all of Hong Kong's imports of industrial roundwood and plywood came from countries within the Asia-Pacific region (FAO 1995).

Hong Kong companies have a long history of offshore investment in forestry, but details of specific companies that are or have been involved, and any impacts that their operations may have caused in host countries, are scarce. Between 1967 and 1989, Hong Kong companies were the biggest investors in the wood product and paper manufacturing industries of Indonesia, accounting for 36.4 percent and 41.6 percent of foreign investment in these sectors over this period, respectively (Hill 1996). Recently, a Hong Kong holding company showed interest in obtaining concessions for most of Gabon's remaining timber resources (Bohan 1996). Another role Hong Kong has played in the tropical timber trade is that front companies situated there have facilitated transfer pricing on log exports and log smuggling from, at least, the Philippines, Malaysia, and Papua New Guinea (Callister 1992).

Logging the Frontier: Case Studies

Four cases from the tropical forest frontier illustrate the problems associated with increased logging and the degradation that can result.

Belize. About 70 percent of Belize is covered by broadleaf tropical forest. Belize has prided itself on a history of environmental protection (Loftis 1996). The government is now seeking much-needed revenues for its low coffers, however, and since the 1980s, has been granting large logging concessions in southern Belize.

Seventeen concessions, totaling about two hundred thousand hectares, have been granted by the Ministry of Natural Resources, sixteen of them since 1993 (Cho 1996; Indian Law Resource Center 1996). Many of the concessions are situated in the Columbia River Forest Reserve, a protected area since 1953. Some concessions almost completely overlap Maya Indian villages and threaten to deprive these groups of their livelihood (Loftis 1996). The timber concessions were negotiated in private sessions with the government. The first publicized concession was granted to Belize Fine Woods Limited in 1994, only to be suspended for faulty management and then abandoned a few months later when the firm's owner filed for bankruptcy. The following year, the same businessman obtained a much larger concession as the negotiator and license holder for Malaysian-owned Atlantic Industries Limited, which was granted a twenty-year concession for mahogany logging of eighty thousand hectares in the Toledo region of southern Belize (Loftis 1996; Ito and Loftus 1997).

The contract was granted over the objections of the government's technical forestry consultant. Logging began in September 1995 and included the construction of a sawmill near a Mayan village. After villagers protested and noted that the zone in question was not supposed to be logged until the year 2007, the company was forced to relocate and began construction of another sawmill, which, when completed, will be the largest in Central America (Cho 1996; Ito and Loftus 1997).

Actual operations are virtually unsupervised, since Belize has only four forestry inspectors to monitor the entire country. Newspaper reports in Belize confirmed that logging companies in the region were violating conditions governing their concessions, including breach of starting dates, consultation with local communities and residents, and logging of protected species (*Reporter* 1996).

Suriname. In 1993, the new government of Suriname sought aid to restructure and build an independent economy. In a country where almost 90 percent of the land area is forest cover, timber concessions seemed an easy solution. By February 1995, at least five foreign investors had sought concessions for timber production that were greater than the maximum size of 150,000 hectares permitted under the 1992 Forest Management Act. Of these, three requests—from Malaysia's Berjaya Group and Indonesia's Mitra Usaha Sejati Abadi (MUSA) and Suri Atlantic—were for over one million hectares each, with total projected investments of around US$300 million.

Some recent incidents cast a shadow over Berjaya's corporate responsibility. In July 1994, for example, the managing director of Berjaya Group Limited in the Solomon Islands was expelled from the country for allegedly attempting to bribe the Solomon Islands' minister of commerce, employment, and trade (*Timber Trades Journal* 1994).

MUSA Indo-Suriname[4] is a logging company that forms part of a larger timber and industrial consortium under Indonesia-based Porodisa Trading. MUSA has operated a 150,000-hectare concession in Apoera in western Suriname since 1993. In obtaining the concession, the company falsely claimed close ties to high-ranking political and business officials in Indonesia and Suriname. MUSA also created at least sixty-nine shadow companies, almost all of which were involved in forestry activities, in order to circumvent parliamentary oversight and attempt to obtain a larger total concession area (totaling eight-million hectares).

MUSA's operation has repeatedly come under fire in the local press and internationally. Logging within the Apoera concession has proceeded without the proper surveys and management plans, and at the same time, MUSA's operators have been dispatched to areas far outside the official concession, provoking widespread environmental destruction. Conflicts have developed between MUSA and a Suriname state company over logging tracts and between MUSA and local communities over compensation for timber and construction. The company has violated labor recruitment and immigration laws and engaged in bribery. After widespread publicity by international environmental groups, the Suriname National Assembly delayed votes on granting the concessions and, in 1996, rejected all applications for the large-scale concessions.

The government of Suriname has very limited capacity to monitor logging operations or to enforce the Forest Management Act of 1992. Furthermore, the economic benefits of expanding logging operations are greatly exaggerated, while the risks of environmental and social impact are substantial (Sizer and Rice 1995).

Following elections in mid-1996, there was a significant change of government in Suriname. In order to govern, a coalition was formed among a number of smaller parties, including groups tied to the foreign logging interests represented by MUSA and the Berjaya Group. Despite these links, the new government publicly committed to put a hold on issuance of the proposed major new logging concessions.

Local nongovernmental organizations (NGOs) reported in late 1997 that Barito Pacific was in the process of acquiring six hundred thousand hectares of concessions in central-east Suriname. According to recent reports, Barito representatives were accompanied by members of the Suriname National Army when visiting areas occupied by Maroon communities. These communities have stated their objection to these and other concessions.

Guyana. One of the most heavily forested countries in the world, Guyana has forest cover over sixteen million of its nineteen million hectares. In order to garner benefit from its vast natural resources, Guyana began in the early 1990s to seriously consider granting logging concessions to Asian and North American companies.

Starting in 1995, donors began work with the government to implement a process of strengthening the government's capacity to monitor logging and to enforce the law. Commitments were also made to update Guyana's laws affecting use of forest resources. As part of the process, the Guyanese agreed to halt the issuance of new large logging licenses. In addition, a new law was to be prepared allowing for "exploratory" licenses, which would permit foreign companies to apply for licenses to study potential concessions, during which time they would be required to pay area fees and application fees with no guarantee of actually getting a license to log (Sizer 1996). The new law was approved in late 1997, and applications are being considered. The changes in the law in Guyana could potentially substantially increase government revenue from logging and timber export and improve monitoring and law enforcement. At the same time, there could be a significant expansion of logging operations, while the country continues to lack even a basic network of protected areas and while several dozen Amerindian land claims remain unresolved.

At least eight foreign loggers have made requests for substantial concessions in recent years (over five hundred thousand hectares each). In October 1991, under Guyana's previous government, the Barama Company Limited won a twenty-five-year, extendable investment contract that gave the company logging rights to 1,690,000 hectares (almost 10 percent of the national territory) in the northwest, near the Venezuelan border. Barama is a wholly owned joint venture between Samling Corporation of Malaysia (80 percent ownership) and Sunkyong Limited of Korea (20 percent). The total proposed investment, according to the company, was US$154 million (Sizer 1996).

Barama is committed to extracting timber from their concession using sustainable forestry practices. On contract to Barama, the Edinburgh Center for Tropical Forests (ECTF) is doing research at the site and is developing a program of environmental monitoring. Barama has purchased logs from other producers, over whose logging standards it has no control, in order to meet orders on time and to keep its plywood mill running.

At least five Amerindian communities live within or on the periphery of the concession. Four are recognized as Amerindian villages under the Amerindian Act. Under Guyana's law, their village lands are excluded from the concession area. Several smaller communities and families are scattered along the Barama River and other rivers (Forte and Pierre 1995).

Estimates based on data provided by the Guyana Forestry Commission and by the company itself show that Barama pays the equivalent of less than 1 percent of the value of the goods exported in taxes (Sizer 1996). It appears that Barama may be getting an unusually good deal. To date, however, financial results have been below initial expectations. Total production of plywood, and especially of logs, has been low, and unit production costs have been high (Sizer 1996).

Papua New Guinea (PNG)

Recent estimates of PNG's forest cover vary from thirty-six million hectares (77 percent of total land area) to thirty-nine million hectares (85 percent of the country) (World Bank 1989; PNGFA 1996a). Until recently, the country's remoteness and the rugged central mountain range have made it less accessible to commercial logging companies. But within the past decade, the export of raw logs by multinational corporations has grown rapidly. The industry is dominated by Malaysian multinational logging companies, led by giant Rimbunan Hijau, selling to Japan and other Asian timber processing countries.

PNG has been experiencing a logging boom. An inevitable bust looms with the recent slowdown in demand from Asian markets. The economic returns will inevitably dwindle as the initial harvest is completed. Within the past ten years, logging in PNG has grown to unprecedented amounts, increasing by over 300 percent in 1993 alone (Filer 1997). Japan is the destination for over half of the logs and South Korea consumes about one-quarter of the raw log exports. There has also been a very rapid growth in exports to the Philippines following exhaustion of their domestic forest resources (Filer 1997).

In 1989, the Barnett Commission of Inquiry into the Timber Industry documented rampant corruption and abuse of privilege at all levels and how such practices devastated the people and the environment. According to the Commission, "[i]t would be fair to say, of some [of the foreign companies], that they are now roaming the country with the self assurance of robber-barons; bribing politicians and leaders, creating social disharmony and ignoring laws and policy" (Barnett 1989).

If current logging trends continue, PNG will have lost the majority of its rain forests within two decades. It is estimated that 84 percent of PNG's forest frontier is threatened by expansion of logging operations. The more detailed impacts of this logging on the environment and local communities has been documented through various thorough case studies summarized by Colin Filer (1997).

One case describes the impacts of logging in the Hawain Local Forest Area. Following several years of tense negotiation and conflict, a company called Sovereign Hill, a subsidiary of Malaysia's Rimbunan Hijau, won permission from local communities to log in this area. The clearance of logging roads along ridges led to rapid soil erosion and siltation of creeks and streams below. As a result, there was a significant decline in the quantities of fish and crustaceans formerly consumed by the local residents. Forest fauna were also severely affected. Villagers reported that the noise of bulldozers and chainsaws caused large game animals to flee the area. Bulldozers destroyed the ground nests of brush turkeys, and the clearance of top ridge forests made it difficult to trap flying foxes. In addition, expatriate and local employees of the logging company also hunted the fauna displaced by the tree felling.

For the customary landowners whose forests were logged, the financial benefits of royalty payments proved to be disappointing and short-lived. While they invested some of the income in used motor vehicles, trade store inventories, or savings accounts, most of it was used to meet customary exchange obligations, dealt out to close kin in small quantities, and spent on sponsoring beer parties. At the peak of the operation, Sovereign Hill employed only fifty to sixty local people, mostly as chainsaw operators, drivers, and assistants. Women's weekly income from selling garden produce in town exceeded the low wages.

Final Thoughts

In industrialized countries there have been substantial efficiency gains that serve to dramatically reduce the demand for wood in the long term. While use of processed products has grown very rapidly, the consumption of raw roundwood inputs has been practically stagnant, rising by only 15 percent in developed countries. Eventually, these efficiency gains will also be shared with developing countries as they increase their competitiveness.

There is a growing disconnect between use and raw material supply. Although global wood pulp production has declined since 1990, this has been more than compensated by a 21 percent increase in waste paper collection for recycling. Furthermore, plantation growth rates are five to ten times those of managed natural or second growth forests, and in some cases far higher, aiding in replacement of supply from natural forests with supply from what are essentially "fiber farms."

A North American company called Earthshell is now making paper and packaging out of starch and limestone. The Brazilian firm Aracruz is making building material out of eucalyptus from its Brazilian plantations. And silicon-based electronic paper is in development. These technological innovations could all dramatically alter demand for wood-based products and in turn help to reduce the pressure on remaining forest frontiers.

Meanwhile, there has been an important shift in the origin of the investment and capital behind the expansion of transnational logging operations in the tropics. A large part of the new investment (at least prior to the recent economic downturn in Asia) is controlled by companies from countries such as Malaysia and Indonesia. Twenty years ago, almost all the investment was from Japan, the United States, and Europe. Many countries are turning over huge areas of primary tropical forest to logging without having first identified and delimited areas of critical importance for biodiversity conservation, watershed management, or the production of other goods and services. Often traditional land rights and indigenous territories issues have not been resolved either.

Lack of good governance is a key part of the problem. Often decisionmaking is controlled by a small group of powerful people or clans within the government who look at primary forests as a short-term source of personal revenue, not as a productive ecosystem that can generate social, economic, and ecological benefits on a long-term basis for the entire country and its people.

A set of specific measures that should be taken before further investment is encouraged would include definition and demarcation of the permanent forest estate with a clear indication of production, protection, and conservation forests. Areas for indigenous peoples communities should be clearly demarcated. Before new logging rights can be awarded, there should be a thorough analysis of priority areas for watershed protection and conservation of biodiversity and other nontimber values, as well as resolution of resource use and land rights disputes with indigenous or tribal communities, who need to be involved in the decisionmaking process from the very beginning. Without such definition, investment will be far less attractive to more responsible investors seeking to reduce the risk of disputes over property rights. A clear indication of the permanent production forest areas is furthermore indispensable for any long-term forest management for timber production.

Governments should ensure that timber-cutting rights are allocated in a transparent and competitive manner that discriminates in favor of investors with a proven commitment to responsible forest management. Investors should be required to pass through a "filter" of assessment that selects those with a proven track record and commitment to responsible practice.

Furthermore, timber extraction should be taxed through simple, easily enforceable means at an appropriate level, and sufficient funds should be invested in increased education, research, capacity-building, and training, as well as in the overall improvement of forest management. The weak contract negotiation, tax assessment, auditing, and collection capacity of many governments has resulted in widespread loss of revenues. Governments should also ensure that they have top quality international legal advice available when they negotiate with foreign investors.

Specific measures should be developed to address the indirect impacts of logging operations such as commercial hunting, fires, and movement of farmers into forested areas along logging roads. Logging companies should be held responsible through legal measures and financial incentives for minimizing the indirect effects of their operations that arise through the creation of access roads and disturbance of the forest ecosystem. Effective checkpoints on access roads, closure of feeder roads that are no longer in use, and the prohibition of commercial hunting can help greatly to reduce the indirect impacts of logging. EIAs for all major investments should be used proactively to identify potential negative impacts and steps necessary to avoid them.

Donors could do more to tie bilateral and multilateral assistance to good governance and a commitment to responsible forest management. Governments that make and stick to commitments to behave responsibly should receive preferential attention from donors and technical assistance agencies. Preferential assistance should be given to those countries that have the courage to enact moratoria on new investments until they have established a national system of protected areas, have addressed indigenous and tribal land claims, and have created sufficient technical capacity to monitor the logging activities.

Acknowledgment Much of the information presented in this chapter, including the data on investment and the four local case studies, is drawn from the European Commission–sponsored joint WRI/WWF report by Sizer and Plouvier (1998).

NOTES

1. *Roundwood production* refers to all wood in the rough, whether destined for industrial or fuelwood uses. All wood felled or har-

vested from forests and trees outside the forest, without bark, round, split, roughly squared, or in other forms such as roots and stumps, is included.

2. The auditor-general of Malaysia has said that "imposing fines on such [illegal loggers] under the existing laws is inadequate, and as such more deterrent punishments should be introduced by the legislative body." Quoted in Chaong and Kuttan (1992).

3. "Transfer pricing of log exports was common, allowing timber merchants to realize most of their profits in privately held offshore companies and reduce their Malaysian tax exposure" (Pura 1994c). "Privately, some timber industry executives and bankers acknowledge that the practice [of transfer pricing] was widespread" (Pura 1994a).

4. This section is a summary of SKEPHI. 1996. *Asian Forestry Incursions: Report on N. V. MUSA Indo-Surinam.* International Fund for Animal Welfare, Brussels.

REFERENCES

Barnett. 1989. "Report of the Commission of Enquiry into Aspects of the Forest Industry: Final Report." Unpublished report to the government of Papua New Guinea.

Bohan, et al. 1996. *Corporate Power, Corruption and the Destruction of the World's Forests.* Environmental Investigation Agency, London, England.

Brown, P. 1988. *Climate, Biodiversity and Forests: Missed Opportunities or Double Dividends.* World Resources Institute, Washington, D.C.

Brunner, J., et al. 1988. "Logging Burma's Forests." Draft. World Resources Institute, Washington, D.C.

Bryant, D., et al. 1997. *The Last Forest Frontiers: Ecosystems and Economies on the Edge.* World Resources Institute, Washington, D.C.

Callick. 1995. An unread message from a rainforest: Logging media and social conflicts. In *Our Trees and All the Wildlife Have Gone*, edited by F. Kürschner-Pelkmann, M. Trott, and I. Wöhlbrand. Association of Protestant Churches and Missions in Germany, Hamburg, Germany.

Callister. 1992. *Illegal Tropical Timber Trade: Asia-Pacific.* WWF-International, Gland, Switzerland.

———. 1996. "Asia-Pacific Wood Product Trade: An Overview and Preliminary Assessment of Implications for WWF's Forest Conservation Activities." Unpublished report for WWF's Forests for Life campaign. Gland, Switzerland.

Carrere. 1996. Pulping the south: Brazil's pulp and paper plantations. *Ecologist* 26(5):206–214.

Cho. 1996. *Massive Logging in Toledo Still Imminent.* Toledo Maya Cultural Council, Belize.

Choong, A., and S. Kuttan. 1992. Illegal logging: State-owned companies biggest culprits. *New Straits Times*, November 25.

Crossley, R. 1996. *A Preliminary Examination of the Economic and Environmental Effects of Log Export Bans.* Environment Department, World Bank, Washington, D.C.

Denny. 1995. Malaysian joint venture forestry deal. *Stabroek News*, November 8, p. 1.

Dudley, N., and S. Stolton. 1994. *The East Asian Timber Trade and its Environmental Implications.* WWF–United Kingdom, Godalming, England.

FAO. 1995. *Yearbook of Forest Products 1982–1993.* United Nations Food and Agriculture Organization, Rome, Italy.

———. 1997. *State of the World's Forests.* United Nations Food and Agriculture Organization, Rome, Italy.

Filer, C., ed. 1997. *The Political Economy of Forest Management in Papua New Guinea.* National Research Institute of Papua New Guinea, monograph no. 32, and the International Institute for Environment and Development, London.

Forrest. 1991. Japanese aid and the environment. *Ecologist* 21(1): 24–32.

Forte, and L. Pierre. 1995. Survey of forest use in region I. In *Situation Analysis Indigenous Use of the Forest with Emphasis on Region 1.* Amerindian Research Unit, University of Guyana, Georgetown, Guyana. Pp. 12–22.

Gray. 1996. Asian logging giants extend reach. *Associated Press,* Kuching, Malaysia, August 18.

Gresham. 1994. Timber trends. *Tropical Forest Update* 4(4).

Hamid. 1993. Cabinet wages all-out war on illegal logging. *Business Times,* January 28.

Hill. 1996. Manufacturing industry. In *The Oil Boom and After: Secret Belize Logging Concessions Threaten Maya Homeland Tropical Forest,* edited by A. Booth. Indian Law Resource Center, Washington, D.C.

Indian Law Resource Center. 1996. *Secret Belize Logging Concessions Threaten Maya Homeland Tropical Forest.* Indian Law Resource Center, Washington, D.C.

Indonesian Economic Policy and Performance in the Soeharto Era. Oxford University Press, Singapore. Pp. 204–257.

Ito, and M. Loftus. 1979. Cutting and dealing: Asian loggers target the world's remaining rain forests. *U.S. News and World Report,* March 10, p. 40.

ITTO. 1996. *Annual Review and Assessment of the World Tropical Timber Situation.* International Tropical Timber Organization, Yokohama, Japan.

Khan, K. M. 1986. "Multinationals from the South: Emergence, Patterns and Issues." In *Multinationals of the South: New Actors in the International Economy,* edited by K. M. Khan. Prances Pinter, London, England. Pp. 1–14.

von Kirchbach. 1983. *Economic Policies towards Transnational Corporations: The Experience of the ASEAN Countries.* Nomos, Baden-Baden, Germany.

Kürschner-Pelkmann. 1995. Wood and paper: Development, ecology and the media in the Asia-Pacific region. In *Our Trees and All the Wildlife Have Gone,* edited by F. Kürschner-Pelkmann, M. Trott, and I. Wöhlbrand. Association of Protestant Churches and Missions in Germany, Hamburg, Germany. Pp. 5–70.

Loftis. 1996. For love of land. *Dallas Morning News,* September 15, p. 1A+.

Lynch. 1992. *Securing Community-Based Tenurial Rights in the Tropical Forests of Asia.* World Resources Institute, Washington, D.C.

Ministry of Forestry, Jakarta. 1994. *Resource Rent and Implications*

for Sustainable Forest Management. Sector Study Working Paper
no. 3, ADB Project Preparation Technical Assistance no. 1781-
INO. Ministry of Forestry, Jakarta, Indonesia.

———. 1993. *Tropical Forests of Indonesia: A Country Paper*. Ministry of Forestry, Jakarta, Indonesia.

Nectoux, and Y. Kuroda. 1989. *Timber from the South Seas*. WWF-
International, Gland, Switzerland.

Newell, and E. Wilson. 1996. The Russian Far East: Foreign direct investment and environmental destruction. *Ecologist* 26(2):68–72.

Nurse. 1995. New forestry leases on the cards. *Guyana Chronicle*,
November 6.

OECF. 1997. *Annual Report 1997*. Overseas Economic Cooperation
Fund, Tokyo, Japan.

Papua New Guinea Forest Authority. 1996a. *Annual Report for
1995*. PNGFA, Boroko, PNG.

———. 1996b. *The National Forest Plan for Papua New Guinea*.
PNGFA, Boroko, PNG.

Payne. 1994. Asia-Australasia limbers up for the twenty-first century. *Pulp and Paper International* 36(6):59–61.

Pura, R. 1994a. Planned Malaysian listing draws attention of tiongs.
Asia Wall Street Journal Weekly, February 21, p. 16.

———. 1994b. Timber baron emerges from the woods. *Asian Wall
Street Journal*, February 15, 1998. p. 1.

———. 1998. Timber companies blossom on Malaysian stock market. *Asia Wall Street Journal*, November 30, p. 12.

Reid, and K. M. Miller. 1989. *Keeping Options Alive*. World Resources Institute, Washington, D.C.

Reporter, Belize. 1996. Maps reveal conflict between Mayas and logging licenses. October 13, p. 4.

Sar, and P. Sledge. 1994. Outlook for Australian woodchips. *Quarterly Forest Product Statistics*, vi–ix.

Schwartz, and J. Friedland. 1992. Green fingers. *Far Eastern Economic Review*, March 12, pp. 42–44.

Sizer, N. 1996. *Profit without Plunder: Reaping Revenue from Guyana's Tropical Forests without Destroying Them*. World Resources Institute, Washington, D.C.

Sizer, N., and D. Plouvier. 1998. *Increased Investment and Trade
by Transnational Logging Companies in Africa, the Caribbean
and the Pacific: Implications for the Sustainable Management
and Conservation of Tropical Forests*. World Resources Institute
and World Wildlife Fund, Washington, D.C.

Sizer, N., and R. Rice. 1995. *Backs to the Wall in Suriname: Forest
Policy in a Country in Crisis*. World Resources Institute, Washington, D.C.

Suharyanto. 1992. Greening Indonesia: Timber cutters come to
party. *Indonesia Business Weekly*, August 13, p. 23.

Timber Trades Journal. 1994. Solomon Islands logging project to go
ahead despite bribery allegations." September 10.

Tsuruoka. 1993. Vanishing coral reefs: Plundering threatens Tioman
and other Asian tourist centers. *Far Eastern Economic Review*,
January 7, pp. 24–25.

Walker, and J. A. Hoesada. 1986. Indonesia: Forestry by decree. *Journal of Forestry* 84(11):38–43.

World Bank. 1989. *The Forestry Sector: A Tropical Forestry Action Plan Review*. World Bank, Washington, D.C.

World Rainforest Movement. 1995. Solomon Is.: Anti-logging protesters detained on Pavuvu Island. *WRM Newsletter* 31, p. 13.

WRI. 1998. *World Resources Report, 1998–99*. World Resources Institute, Washington, D.C.

8 Aracruz Celulose

A Case History

Erling Lorentzen

Common perceptions of forestry companies in South America tend to brand them with the image of decades of rain forest destruction. In some areas, where extraction of prized hardwoods has been the dominant industry, that reputation has historically been deserved, with uncontrolled logging removing whole swathes of native forest to the detriment of biodiversity and local communities. However, governments and NGOs alike have increasingly come to realize that better control of logging concessions is only part of the answer to conserving the biodiversity of regions' native forests. Poverty is, in most cases, a far more dangerous enemy than the search for corporate profit. Extraction of timber by impoverished local communities, either for fuel, construction, or simply land clearance for planting, has been estimated, for example, to account for the considerable majority of native forest loss in some areas of the Amazon.

One approach to the problem over the past thirty years is for government to work closely with forest products companies to reduce pressures on dwindling native forests—first, by providing a new economic framework in the regions most affected so that the damaging effects of poverty on the native forest are mitigated; second, by developing alternative, *planted* sources for timber and cellulose that make extraction from the native forest unnecessary.

In this context, the Brazilian government approved a project in the state of Espírito Santo in the mid-1960s, at a time when the Atlantic rain forest that originally covered almost 90 percent of the state had shrunk considerably. Much of the southern part of the state had been laid waste by successive cycles of logging, agriculture, cattle-raising, and charcoal-burning. The new project was an innovative one for Brazil at the time: creation of planted forestry, using fast-growing eucalyptus, for the manufacture of pulp for the world's paper industry. The company established to implement the project was Aracruz Celulose.

Pioneering

Since the start of the project in the late 1960s, the necessary private investment to develop the required infrastructure—roads, housing, schools, health care facilities—has absorbed some US$125 million out of a total investment to date of over US$3 billion. Meanwhile, Aracruz's forestry operations have grown to cover 203,000 hectares in Espírito Santo and the neighboring state of Bahia. Lands owned today include 132,000 hectares of eucalyptus plantations, interspersed with native forest reserves and other original ecosystems to encourage biodiversity. The company's associated pulp mill, adjoining its plantations on the coast of Espírito Santo, has an annual capacity of 1,240,000 tonnes, making it currently the largest single mill in the world. The company's adjacent private port facility, Portocel, is the largest specialist pulp-exporting port in Brazil.

One of the keys to the company's commercial success has been the pioneering development of high-yielding eucalyptus cloning techniques. Coupled with Brazil's natural climatic advantages, Aracruz has been able to create one of the shortest tree harvest cycles in the world: only seven years from planting to maturity. The remarkable productivity of eucalyptus is significantly higher than any other species used for pulp production. For example, wood yields from Aracruz's plantations are a world-leading forty-four cubic meters per hectare per year, ten times greater than the productivity normally achieved in countries of the Northern Hemisphere. Such technological advances have made the company one of the lowest cost producers in the world.

Table 8.1 shows the remarkable efficiency of typical high-productivity eucalyptus in Brazil, compared with other tree crops in different producing countries.

Through this combination of natural and manmade advantages, Aracruz has become, over the last thirty years, the world's leading supplier of bleached eucalyptus market pulp—mostly used in the production of high quality tissue and fine papers—and accounts for 22 percent of the worldwide demand for this type of pulp. In addition, Tecflor Industrial, a joint venture company, from early 1999 will be supplying high-quality timber from Aracruz's plantations to the decorative millwork market for products such as flooring, windows, door components, and furniture.

Figure 8.1   Brazil.

Aside from their undoubted commercial potential, can such large-scale forestry operations make a positive contribution to conserving biodiversity, or are they simply a way of slowing the rate of its loss? And do they confer genuine benefits on local communities, including indigenous peoples, or do the profits mostly wind up in the hands of entrepreneurs and investors? The answers depend to a large extent on the way such forestry operations are established, the philosophy behind their operations, and the manner in which they are run. Perhaps, in an increasingly cynical world, it also depends on how easily

Figure 8.2   Planted Forestry.

Table 8.1 Rotation Age and Average Productivity of Forest Species Used for Pulp Production

Species	Country	Rotation (Years)	Mean Increment (M³/ha/year)
Hardwood			
Eucalyptus hybrid	Brazil (Aracruz)	7	44.0
Eucalyptus grandis	South Africa	8–10	20.0
Eucalyptus globulus	Chile	10–12	20.0
Eucalyptus globulus	Portugal	8–10	12.0
Eucalyptus globulus	Spain	12–15	10.0
Betula spp.	Sweden	35–40	5.5
Betula spp.	Finland	35–40	4.0
Softwood			
Pinus radiata	Chile	25	22.0
Pinus radiata	New Zealand	25	22.0
Pinus spp.	Brazil	15–20	16.0
Southern pines	United States	25	10.0
Douglas fir	Canada (BC coast)	45	6.6
Norway spruce	Sweden	70–80	4.0
Norway spruce	Finland	70–80	3.6
White spruce	Canada (BC inland)	55	2.5
Black spruce	Canada (east)	90	2.0

Source: Jaakko Pöyry. 1994. Portucel. 1995.

they can be called to account and how responsive they are to challenge on their environmental and social records.

Sustainable Development

In the case of Aracruz Celulose, the project was based on the principles of sustainable development—insofar as they were understood at the time—right from its inception in the 1960s. Pulp has always been produced from planted forestry, and native trees have never been exploited for pulp manufacture or for any other reason. Likewise, the company's pulp mill uses energy-saving biomass from its tree debarking process to generate electricity, and Aracruz has invested in advanced systems to treat all solid, liquid, and gaseous wastes.

In the interests of ecological balance and biodiversity, Aracruz's 132,000 hectares of eucalyptus plantations are interspersed with 57,000 hectares of conservation areas. These areas, which account for 28 percent of the company's total landholdings, are permanently protected and contain a wide range of ecosystems, including salt marshes, mangroves, tableland forest, secondary forest, and swamplands. In order to enrich them, Aracruz maintains a program specifically to produce native seedlings for planting. During the past six years alone, 4.6 million seedlings from 250 separate species native to the Atlantic rain forest and its associated ecosystems have been produced. These include the peroba-amarela (*Paratecoma peroba*), sucupira (*Bowdichia virgilioides*), jacarandá (*Dalbergia nigra*), boleira (*Joannesia princeps*), pink jequitibá (*Cariniana legalis*), cedar (*Cedrela fissilis, Cedrela odorata*), jatobá (*Hymenaea aurea*), gonçalo-alves (*Astronium graveolens*), and ipê (*Tabebuia* spp.), among others. By 2003, it is estimated that a further four thousand five hundred hectares of Aracruz lands will be similarly enriched with native species to improve the biological balance of the forests.

In a separate initiative, Aracruz, in partnership with federal and state institutions, has developed the Forest Rehabilitation and Preservation Program, under which a further two million seedlings of various native species have been distributed from the Aracruz nursery to small rural producers. The resultant plantings serve both to encourage gradual reinstatement of areas of native forest and to reduce the pressures on the small amounts that still survive.

The result is that contemporary satellite images of Espírito Santo now show a marked contrast between the green of Aracruz's plantations and native reserves and the surrounding areas of poorly maintained and largely deforested land. The areas of native forest preserved within Aracruz's properties are, in fact, one of the last areas of Atlantic rain forest remaining in the state, the total land area of which is now only 8 percent native forest. Table 8.2 shows the relationship

Table 8.2 Land Use in Espírito Santo
(Total territory: 45,733 km²)

Pastures	41.2%
Others	18.2%
Permanent crops	15.7%
Temporary crops	9.3%
Native forests	8.8%
Plantation forests	3.4%
Nonutilized productive areas	3.4%

Source: IBGE. 1985.

that exists today between the planted forestry in the state and other uses of the land area.

The benefits of the Aracruz project to biodiversity in the region are considerable; the Aracruz forest reservations have one of the highest biodiversity indices in the world (rating five on the Shannon-Wiener scale). Over 460 tree species per hectare have been identified on the tablelands.

The presence of a very diversified flora also provides a habitat for wildlife. Some 1,703 animal species have been identified on the company's land, including eighteen threatened with extinction. These include the masked titi monkey, ocelot, margay, mantled hawk, bushmaster, and yellow-throat cayman.

The most successful environmental initiatives generally reveal a mixture of altruism and self-interest, and Aracruz's efforts to maintain biodiversity in its areas have had the additional benefit of proving highly effective in natural pest control. The improved environmental balance has been found to provide a barrier against insect attacks in the plantations, with biological control agents in the preserved native forest areas having marked benefits on the neighboring eucalyptus. Such agents include birds, predatory insects, and other microorganisms that are natural enemies to tree pests. For example,

Table 8.3 Fauna Identified on Aracruz's Land

	Families	Species	Threatened Species
Mammals	24	65	10
Birds	56	444	18
Reptiles	18	54	1
Amphibians	6	47	—
Fish	34	98	—
Crustaceans	10	28	—
Total	148	736	29

Source: Aracruz Celulose. S.A. 19XX.

Table 8.4 Wood Use in Brazil

Type of Demand	Million M³	Percentage
Firewood	139.50	49.4
Charcoal	98.40	34.8
Pulp	18.00	6.4
Lumber	17.40	6.2
Panels	3.71	1.3
Chipboard	2.12	0.7
Others	3.18	1.2
Total	282.31	100.0

Source: Brazilian Society of Silviculture. 1993.

human intervention in the form of biological insecticides to control leaf-eating caterpillars (a common forestry problem) has been necessary in only 0.3 percent of infestations in the plantations over the past few years.

A further aspect of Aracruz's operations has helped not only to reduce the pressures on native forest but also to extend the social benefits of its operations in Espírito Santo to the neighboring state of Minas Gerais. In 1991, the company established the Partners in Timber program, designed both to provide an additional source of income to small local farmers and to help reduce pressures on the native forest. Under the program, the company supplies eucalyptus seedlings and technical help to some two thousand farmers in Espírito Santo and Minas Gerais. The seedlings are used to establish plantations on the farmers' own small holdings; the resultant tree crops provide the farmers with an additional income source when the trees are sold to Aracruz for pulping plus a wood supply for their own uses. To date, some nineteen thousand hectares have been planted under this program, yielding over ninety thousand cubic meters of wood annually—around 5 percent of Aracruz's pulp wood requirements. Other observable benefits include additional incentives for farm workers to stay in the rural areas, gradual recovery of degraded land areas, and promotion of environmental education in the local communities.

The importance of such a program is illustrated by the above chart, which shows that the biggest threats to the native forest in Brazil come from domestic requirements such as firewood and industrial ones such as charcoal.

Of the current wood consumption in Brazil of 282 million cubic meters per year, only 75 million is supplied by wood from planted forests. The remainder—estimated at 207 million cubic meters—comes from native tropical forests. As table 8.4 shows, nearly 85 percent is used for energy purposes. In contrast, all of the pulp produced in Brazil comes exclusively from planted trees.

Today, in addition to Aracruz's operations, eucalyptus plantations are found in almost all Brazilian states, including Minas Gerais, São Paulo, Espírito Santo, Bahia, Rio Grande do Sul, Paraná, Pará, Mato Grosso, Rio de Janeiro, and Goiás. They cover an estimated area of three million hectares, or 0.35 percent of Brazil's territory. The amount of these plantations dedicated specifically to paper and pulp production corresponds to 0.1 percent of the nation's territory.

Social Benefits

The social benefits to the communities in which Aracruz operates come in the form both of inputs to the national and local economies—which are easy to quantify—and improvements in the local quality of life, which are more intangible but equally real. From 1989 to 1997, the company generated over US$4 billion of wealth, of which more than half was reinvested in the business, much of it in the region of its operations. In addition, an average of over US$200 million a year has been injected into the local economy through local salaries, state taxes, community projects, and local purchasing of products and services. This injection has been a key factor in the rise of per capita gross domestic product of Espírito Santo state from 71.3 percent of the Brazilian national average to 92.2 percent over the past twenty years. Annual exports from the state are now almost US$2.7 billion, of which Aracruz's exports constitute 27 percent (US$750 million).

The nature of Brazilian domestic economic policy means that companies generally shoulder a higher measure of responsibility for developing their own infrastructure and providing benefits to their employees than is usual in most North American or European countries. Hence Aracruz's investment of over US$125 million to provide the transport, housing, medical, and educational facilities for its six thousand employees and contractors and their eighteen thousand dependents. Additional voluntary assistance to employees and dependents in the form of vocational training, meals, health care, retirement plans, and transportation typically accounts for US$20 million a year, nearly a fifth of Aracruz's total payroll costs. This figure represents a far higher proportion than in many other countries but is in line with the company's stated aim to be one of the best employers in Brazil.

In the wider community context, Aracruz's community action program has a clear focus. The company avoids paternalistic donations in favor of supporting projects—generally in association with local, state, and federal authorities and nongovernmental organizations (NGOs)—that lead to self-sustaining improvements in local communities' quality of life. In 1997, Aracruz invested US$3.74 million in this way in projects including a new school, an agricultural project to support local Indian communities, an environmental education program, an elementary school teacher training program, and a drug use prevention campaign.

One controversial issue with which the company had to deal during 1998 was a campaign by two local Indian groups, the Tupinikin and Guaraní, to claim lands bought by Aracruz during the 1960s. In spite of its convictions about its legal land rights, Aracruz agreed on a compromise solution with the Indian groups that will result in financial support of US$10 million over the next twenty years for a series of projects designed to help the Indians sustain an improved standard of living. Under the terms of the agreement, the indigenous communities recognized the legitimacy of a decision by the Brazilian ministry of justice, which increased the Indians' reservation area from four thousand five hundred to around seven thousand hectares. Most of the expansion was on Aracruz-owned lands.

One innovative aspect of the agreement is that Aracruz has agreed to develop a partners-in-timber project similar to that already established with small farmers in the state. Beginning this year, the company will be supplying seedlings and technical support to the Indian communities to enable them to cultivate their own eucalyptus in the expanded reservation.

Environmental Management

Monitoring and management of the environmental impact of Aracruz's operations is a complex task, involving the impacts of 132,000 hectares of planted forestry in two states, 57,000 hectares of conservation areas, two thousand independent growers, numerous subcontractors, and the world's biggest pulp mill.

Aracruz's environmental management system is being consolidated in accordance with International Standards Organization (ISO) 14001 principles. Completion of this system is planned for the end of 1998. The system formalizes, in a structured way, several environmental activities that have already been undertaken by the company. In addition, it integrates the environmental activities with quality and health and safety systems, in line with modern management trends. All the company's operations are licensed by the states of Espírito Santo and Bahia.

As part of its modernization program in 1997, Aracruz made further environmental investments of US$62 million. In fact, since its inception, the company has invested a total of US$300 million in environmental control equipment and programs. Currently, expenditure on compliance monitoring, performance improvement, and environmental management systems is over US$2.6 million a year.

In line with a growing number of large corporations operating in environmentally sensitive areas, Aracruz now includes a social and environmental report in its annual report to shareholders, detailing its performance on key indicators of the success of its community and environmental policies. Current data show that the company's operations are well within the compliance levels for all state, national, and international environmental standards.

One interesting initiative is the instigation in 1993 of the Watershed Project, devised by company staff in association with several Brazilian and international institutions. Under the project, a closed site of 286 hectares has been established; it will be monitored throughout an entire seven-year eucalyptus growing cycle. The intention is to gather valuable scientific data on the interaction between the plantation area and the native forest reserves. The complete hydrological cycle is being studied, including water quality, soil fertility, nutrient cycles, bird and insect interactions, impact on local fauna, and the effects of weather conditions on forest growth. The project will yield a unique body of evidence on the interaction between planted and native forest and provide valuable pointers for future forest and water resources management techniques.

New Challenges

The history of any successful business is a story of new challenges encountered and overcome. Likewise, the most valued capability of today's more environmentally conscious boards of directors is their ability to anticipate from where the next challenge will come and to plan how to meet it. For a company doing business worldwide like Aracruz Celulose, challenges can come from anywhere on the globe, such as new ideas on forest management and biodiversity; fresh scientific focus on chemical processes or emissions; controversy over indigenous people's rights; or political impositions in areas such as paper recycling or product ecolabeling.

For example, during the 1980s, controversy built in North America and Europe over the use of chlorine in bleaching pulp for the paper industry, a practice that was believed to result in unsafe levels of dioxin in waste water emissions. Responding to that particular issue involved a major investment in changing Aracruz's bleaching lines, with the result that 67 percent of the company's pulp is now produced without the use of elemental chlorine.

Similarly, in the company's forestry operations, the Watershed Project is a response to uninformed speculation about the effects of eucalyptus plantations on the local water table and ambient native forest.

Another challenge is how to combine a recognition of social and environmental responsibilities with the need to remain competitive in an increasingly globalized marketplace. This will call for the development of new approaches to ecoefficiency and social efficiency similar to productivity strategies deployed elsewhere in the industry.

One thing is certain: in this industry, the winners of the twenty-first century will share several marked characteristics. They will be able to rise to the challenge of meeting the growing demand for cellulose and wood products without adverse effects on forest biodiversity and the environment. They will be able, at the same time, to maintain low costs in their operations. And, equally important, they will be adept at clearly *demonstrating* their environmental and social performance to their customers, investors, employees, and other stakeholders.

9 Stewardship of Mexico's Community Forests

Expanding Market and Policy Opportunities for Conservation and Rural Development

Justin R. Ward
Yurij Bihun

The forests of Mexico have long been recognized as among the world's most unique and biologically diverse ecosystems. However, unchecked deforestation and destructive management practices are degrading wildlife habitat, soil productivity, water quality, and other environmental values that are critical to Mexico's natural heritage and economic future.

Over the last several decades, an alternative model based on environmentally sound community forest management has taken hold in Mexico. A complementary development in the 1990s has been the emergence of independent certification of well-managed forests and labeling of products from certified lands, under the rules of the Forest Stewardship Council (FSC).

This is a time of rapid change in the Mexican forest sector. Recent developments include: major reforms of laws and regulations; financial incentives for plantations and other commercial forest management; an influx of private investment in timber production; international financing of community forest development projects; and growing worldwide consumer preferences for socially and environmentally responsible forest products. Decisions affecting the shape of these changes will have profound, lasting effects on forest landscapes and communities throughout Mexico.

This chapter begins with a snapshot of current conditions and trends in Mexico's forests, with a focus on land ownership, management and production patterns, the ecological importance of the resource, and threats facing forests and forest communities. We then examine the social and environmental benefits of well-managed com-

Figure 9.1   Mexico.

munity forestry in Mexico and the role of FSC-endorsed certification.
The chapter concludes with an overview of some key leverage points
for creating favorable policy climates and market conditions for independently certified forest management in Mexico.

Conditions and Trends in Mexico's Forests

By most current estimates, forests cover approximately fifty million
hectares in Mexico, representing one-fourth of the country's entire
land area (World Forest Institute 1994). Mexico's forested area is di-
vided roughly evenly between tropical forest types and temperate forests.

Mexico's tropical forests, located principally in the southern states
of Yucatan, Campeche, Quintana Roo, Tabasco, Chiapas, Oaxaca, and
Veracruz, represent the northernmost extension of tropical forests in
the Americas. Mexico's temperate forests are characterized by a
"mixed wood" pine-oak complex rich in arboreal diversity. Mexico

146   Forests under Pressure

Figure 9.2   Deforestation.

contains a remarkable climatic complexity and biotic richness that make it one of the world's foremost "megadiversity" countries (Bowles and Prickett 1994). Mexico's biota is noteworthy for its high levels of endemism, meaning that large proportions of species are indigenous to that country.

Forests account for a very substantial portion of the extraordinary species numbers and endemism found in Mexico. With 55 species of pine, of which 85 percent are endemic, Mexico has the world's greatest diversity of the genus *Pinus* (Perry 1991). According to a recent World Resources Institute (WRI) study, Mexico ranks seventh in the world in terms of plant biodiversity in its frontier forest areas, which consist of remaining large, intact natural forest ecosystems that are relatively undisturbed and sufficiently large to maintain all of their biodiversity (Bryant et al. 1997).

Deforestation is a serious problem in Mexico. According to recent data from the United Nations Food and Agriculture Organization, Mexico's forest cover declined an average of 0.9 percent annually during the period 1990–1995; for purposes of comparison, Mexico's rate of forest loss is somewhat lower than the estimated rate of 1.2 percent per year for the Central American region (with Mexico included in the regional calculation) but somewhat higher than the estimated annual rate of 0.5 percent for South America (FAO 1997). There is considerable disagreement over exact deforestation rates in Mexico, with estimates ranging from approximately 300,000 to 1.5 million hectares annually (World Bank 1995a). Mexico has lost 92 percent of its original frontier forest, as defined in the WRI analysis, and 77 percent of

Figure 9.3   Mexican forest.

the country's remaining frontier forest areas are threatened (Bryant et al. 1997).

Land conversion to agricultural cultivation or livestock grazing is a leading cause of deforestation in Mexico, particularly in the southern, tropical region of the country (Clifford 1997). Illegal logging and forest fires also account for significant deforestation in Mexico, especially in temperate forest zones.

The loss of Mexico's forests destroys critical habitat and places species at risk of extinction. The loss of standing forest cover also removes a valuable carbon storehouse that plays an important role in mitigating the atmospheric greenhouse effect and global warming.

Deforestation also causes widespread soil and water degradation throughout Mexico.

Mexico's forest lands are managed predominantly under communal systems of use rights and ownership. Approximately 80 percent of the country's forests belong to some eight thousand forest *comunidades* or *ejidos* located throughout Mexico (Snook 1997). Most of Mexico's forestry communities have high indigenous populations. Individual community land holdings range from one hundred to one hundred thousand hectares.

Mexico has a greater proportion of forest lands under community-based management than any other country in the world. Forest communities account for approximately 40 percent of Mexico's primary wood production and 15 percent of the country's industrial processing of timber (Bray 1997). An estimated eighteen million rural residents depend on Mexico's forests as a source of subsistence and employment (Snook 1997). Mexico is virtually unique among developing countries in having a vigorous community forest movement, notwithstanding that, to date, only a small percentage of Mexico's community forest enterprises have become commercially competitive.

Despite Mexico's generous natural endowment of productive forest lands, the country produces relatively low volumes of wood for commercial purposes. Mexico's timber industry accounts for less than 2 percent of the country's gross domestic product and less than 5 percent of total wood production in North America (FAO 1997). Domestic fuelwood consumption accounts for nearly five times the annual volume of wood harvested for commercial purposes in Mexico (World Bank 1995a).

The Mexican wood products industry is characterized by inefficiencies and lacks economies of scale. Poor road infrastructure complicates timber harvesting and transportation. Ejido sawmills typically operate with outmoded or poorly maintained equipment. The productivity of the labor force is hampered by inadequate training and supervision, as well as frequent work interruptions caused by mechanical breakdowns or loss of workers during peak agricultural planting or harvesting seasons. Very little timber grading and quality control takes place within primary and secondary manufacturing. Lack of marketing and poor utilization of small-diameter, low-grade timber is prevalent.

As a consequence of these and other inefficiencies, the Mexican timber industry is not competitive in international markets. On average, wood produced in Mexico costs 10 to 30 percent more than equivalent wood from the United States or Canada (World Bank 1995a). The price disadvantage of Mexican forest products in North America is compounded by the subsidies, in forms such as below-cost timber sales and artificially low stumpage prices, that are available to some timber producers in the United States and Canada. Mexico has historically been a net importer of wood products, especially pulp and paper, and has exported relatively small quantities of raw or finished wood products, primarily tropical hardwoods, to the United States and Canada.

Social and Environmental Benefits of Well-Managed Community Forestry in Mexico and the Role of FSC-Endorsed Certification

Community forestry in Mexico represents "an experiment in democratic development that can also lead to better and more sustainable environmental management" (Galletti 1998). Community involvement and planning has enabled environmentally sound land use choices with a minimum of conflict over natural resource management (Chapela and Lara 1996). Factors that have been shown to make the difference between success and failure include effective rules to govern management of common property, as well as democratic and equitable distribution of power (Klooster 1997).

During the 1990s, independent certification of forests and forest products emerged as a promising market-based strategy for conservation. The FSC has gained international recognition as the most credible and broadly supported forest certification effort in existence today. By enhancing market opportunities for well-managed community forest enterprises in Mexico, FSC-endorsed certification and labeling could help combat forest destruction and rural poverty.

International Overview of the FSC

Founded in 1993, the Forest Stewardship Council (FSC) is an independent, nonprofit organization encompassing a diverse mix of representatives from environmental groups, the timber trade, the forestry profession, indigenous peoples' organizations, community forestry groups, and certification organizations. The organization sets international guidelines for what constitutes "well-managed" forestry and accredits and monitors certification organizations that evaluate compliance with FSC rules governing on-the-ground forest practices, as well as "chain of custody" tracking systems to ensure that wood products carrying the FSC logo originated from certified forest lands (Upton and Bass 1995).

The FSC Principles and Criteria for Forest Management define in fairly broad terms what FSC-accredited certifiers must evaluate to determine that a managed forest satisfies the requirements for FSC endorsement. The FSC Principles and Criteria are applied in the field through more specific forest management standards covering environmental, social, and economic considerations.

Of particular importance to the environment is FSC principle no. 6, which states that "[f]orest management shall conserve biological diversity and its associated values, water resources, soils, and unique and fragile ecosystems and landscapes, and, by doing so, maintain the ecological functions and the integrity of the forest." Among other requirements, the criteria accompanying principle no. 6 specify that "[c]onservation zones and protection areas shall be established" for endangered species, and that "[r]epresentative samples of existing ecosystems within the landscape shall be protected in their natural state and recorded on maps."

The FSC has undergone steady and dramatic growth during the organization's first five years of existence. There are currently seven FSC-accredited certifiers, each operating internationally. These include SGS Forestry's QUALIFOR Programme and Soil Association's Responsible Forestry Programme in the United Kingdom; the Scientific Certification Systems Forest Conservation Program and the Rainforest Alliance SmartWood Program in the United States; Skal in the Netherlands; Institut für Marktokölogie (IMO) in Switzerland; and Silva Forest Foundation in Canada. As of September 1998, more than 180 forest management certificates, covering a total of more than 17.3 million hectares spread throughout 30 different countries, had been issued by FSC-accredited certifying organizations. The FSC membership has grown to include over 300 groups and individuals from more than forty-five countries around the world.

FSC-Endorsed Certification in Mexico

In Mexico, there is a rapidly growing movement for development and promotion of FSC-endorsed certification. An FSC national initiative is working to raise the public profile of FSC in Mexico and to develop forest management standards that will guide future certification activity throughout the country. Because of the wide diversity of forest types found in Mexico, the standard-setting process has been subdivided into several regions, each with different biogeographic conditions.

Mexico has an emerging "in-country" infrastructure for independent forest certification. The Consejo Civil Mexicano para la Silvicultura Sostenible (CCMSS) is a nonprofit association made up of individuals and nongovernmental organizations (NGOs) with interest and experience in the conservation and management of Mexico's forest resources (Merino 1997). The CCMSS is a founding member of the FSC and currently belongs to the SmartWood Network, which consists of nonprofit organizations throughout the world that collaborate with the Rainforest Alliance's SmartWood Program. As a SmartWood Network member, the CCMSS offers certification services for forest management and chain-of-custody. As the FSC-accredited certification body, the SmartWood Program formally issues, and takes legal responsibility for, certificates that enable marketing claims and labeling using the FSC name and trademark.

Costs of certification vary according to factors such as travel expenses and size of the forest area being evaluated. Generally speaking, the larger the area, the lower the costs per hectare. Certifiers typically calculate the cost of a certification as the sum of the initial evaluation plus the annual audits over the five-year lifetime of the certificate. Measured this way, the costs to the certified forests in Mexico have been in the approximate range of fifteen to ninety-nine cents per hectare (Beyer 1998).

Several community forest operations in Mexico have obtained FSC-endorsed certification. Two noteworthy examples are the four ejidos within the community forestry association Sociedad de Productores Forestales Ejidales de Quintana Roo (SPFEQR—formerly under the Plan Piloto Forestal) managing eighty-six thousand hectares on the Yucatan Peninsula, as well as the four communities that make up the Unión de Comunidades Forestales Zapoteco-Chinanteca (UZACHI), managing twenty-four thousand hectares in Oaxaca. These cases illustrate the community and environmental benefits of FSC-endorsed forest management in Mexico's tropical and temperate regions, as follows.

Certified Forest Ejidos in Quintana Roo

In 1983, the Plan Piloto Forestal, a government-sponsored initiative, began to take shape in the state of Quintana Roo, located on Mexico's Yucatan Peninsula. The project originated during a period of change in which forest use rights were being transferred from the concesionaires to the ejidatarios. Under an agreement with the Mexican government, referred to as the Acuerdo México-Alemania, and funding from the German Development Agency, known as Duetsche Gesellschaft für Technische Zusammenarbeit (GTZ), the project's main objective was to add value to forest resources for the long-term benefit of local communities (Snook 1998). The project was also created as a response to the escalating tropical deforestation brought about by government policies in the 1950s and 1960s that had promoted agricultural conversion and unsustainable timber extraction dominated by the high-grading of commercial species.

Under the Plan Piloto, ten ejidos with the largest and most valuable forests were organized into an informal forestry cooperative that provided technical assistance to the ejidos as well as institutional capacity to negotiate better prices for standing timber. In 1986, these ejidos were officially organized under SPFEQR, a legally constituted association (Kiernan and Freese 1997). The SPFEQR structure proved successful, and by 1992 the model had been expanded in Quintana Roo to four additional associations covering a total of fifty-one ejidos with a combined permanent forest area of roughly five hundred thousand hectares (Argüelles 1993).

In 1991, SmartWood completed its first certification of three ejidal operations belonging to the SPFEQR cooperative. This represented the first independent certification of managed forest land in Latin America. Also in 1991, Scientific Certification Systems (SCS) certified the neighboring Noh Bec and Tres Garantías ejidos. In 1994, SmartWood and the CCMSS worked jointly on a reassessment that culminated in certification contracts being completed with four SPFEQR ejidos: Noh Bec, Caoba, Tres Garantías, and Petcacab. Among these four ejidos, Noh Bec produces the greatest quantities of wood and nontimber forest products. In 1998, Noh Bec withdrew from the SPFEQR cooperative to develop its own office of forest management. This action had no effect on the ejido's certified status. Noh Bec has entered into a technical cooperation agreement with Tropica Rural Latinoamericana, an NGO whose staff had previously been centrally involved in the implementation of the Plan Piloto Forestal.

Within the certified community forests in Quintana Roo, biodiversity conservation is achieved through the identification of large-scale "permanent forest areas" that make up the land base available for timber extraction with environmental safeguards but are not eligible for conversion to pasture or other nonforest use (Galletti 1998). The five hundred thousand hectares of ejido-managed permanent forest area in Quintana Roo provide buffers to, and may function as a biological corridor between, the Sian Ka'an and Calakmul biosphere reserves (Kiernan and Freese 1997).

The timber management systems employed in the permanent forest areas have been shown to maintain high levels of species diversity. Recently, for instance, biologists conducting research in the tropical forests of Quintana Roo concluded that "the type of selective logging done throughout most of the region appears to have no discernible influence on Neotropical migrant birds or on most resident species" (Whigham et al. 1998).

Forest stewardship is practiced at the ejido level, and harvest plans are supervised by a team of local foresters who work for the cooperative and are licensed by Mexico's federal government. Under the new organization of community lands, the ejidos control production, including the harvesting and selling of roundwood and milled lumber. A selection harvest system is planned around the extraction of the two most commercially important species: bigleaf mahogany (*Swietenia macrophylla*) and Spanish cedar (*Cedrela odorata*). The forest management plan for each ejido is designed to assure a "sustained" or continuous yield of mahogany and other species by controlling the rate and spatial distribution of harvesting. Trees meeting a minimum diameter limit are removed under a cutting schedule in which an area the size of one twenty-fifth of the total managed forest is entered in the first year, a different one twenty-fifth of the forest in the second year, and so forth through the end of a twenty-five-year period, at which time the cycle is repeated, with logging of trees that have grown to reach commercial diameters during the first twenty-five-year cycle.

There are indications that silvicultural changes by the certified communities will be necessary to achieve long-term sustained yield of mahogany, a species whose regeneration depends on an ample source of seeds being allowed to grow in unshaded conditions free of competing vegetation. The challenge is to find ways to maintain commercial production of mahogany and to conserve biodiversity. Among the options for creating forest site conditions conducive to mahogany regeneration and growth are more intensive utilization of lesser known tree species and nontimber products, as well as carefully planned shifting agriculture; moreover, sustaining the health of the regional ecosystem and economy will require a diverse mix of timber age and size classes, including the retention of some older mahogany trees left over from logging operations conducted over a period of centuries in Quintana Roo (Snook 1998).

Management adjustments to ensure long-term sustained yield would reduce the annual volumes of mahogany produced. In response to forest inventory data, the SPFEQR communities reduced mahogany production volumes by 54 percent over a recent ten-year period, from approximately thirteen thousand cubic meters in 1984 to approximately six thousand cubic meters in 1993 (Kiernan and Freese 1997).

In light of current and anticipated declines in mahogany volumes, one of the long-term management objectives within the SPFEQR and other community forestry associations in Quintana Roo is to find profitable markets for the region's lesser-known tree species. Lesser known species such as chechen, or "rosewood" (*Metopium brownei*), machiche (*Lonchocarpus castilloi*), chakte-kok (*Sickingia salvdorensis*), and katalox (*Swartzia cubensis*) are developing name recognition in the North American market and are being marketed by some FSC-certified timber distributors. The associations have increasingly been working to incorporate nontimber forest products such as chicle, honey, and wildlife into local forest management and marketing strategies. The FSC is taking steps to integrate nontimber forest products more fully into FSC-endorsed certification and product labeling, which to date has focused predominantly on wood.

Ejido forestry in Quintana Roo has significantly boosted incomes and employment opportunities for many local residents (Galletti 1998). Within each of the four certified communities, at least 50 percent of the ejidatarios are employed in forest management and production (Kiernan and Freese 1997). The associations have channeled substantial income from timber extraction and industrial processing back to community health and education needs and have provided an influential model for state and national policies affecting forests (Silva 1994). The four communities that have obtained FSC-endorsed certification have established market linkages with some businesses interested in wood products from well-managed forests. For example, Gibson Guitar Company is using FSC-endorsed mahogany and chechen harvested from the certified forests in Quintana Roo in production of its Les Paul solid body electric guitars.

Nevertheless, serious economic challenges confront the certified communities and other ejido forestry operations in Quintana Roo. One major constraint is the lack of skills and technology available locally for milling lesser-known species (Kiernan and Freese 1997).

At present, self-financing of environmentally sound forest management is not assured, and few sources of external financial assistance are available to help communities compensate for higher costs or lower revenues associated with conservation-oriented forest management (Galletti 1998). The mixing of community service and business functions within the ejidos' governance structures sometimes results in cash flow problems and inadequate capital reserves within the community forest enterprises (Zabin and Taylor 1997). Notwithstanding limited success in some niche markets, the certified ejidos have not yet realized the dramatically expanded market opportunities they have hoped to achieve within the international certified products trade.

The ejidos' ability to surmount these challenges will hinge on improved capabilities in forest products utilization and marketing. A number of nongovernmental efforts have begun to address this critical need. For example, the Unión Nacional de Organizaciones de Forestería Comunal (UNOFOC), an association of community forestry associations throughout Mexico, has undertaken a commercial promotion effort that links buyers and sellers of timber products (Zabin and Taylor 1997). In partnership with the Tropical Forest Management Trust, based in Gainesville, Florida, UNOFOC is conducting a project known as Mujeres Artesanas, in which women artisans from the Noh Bec community have established a woodturning workshop to make and sell bowls, cups, plates, and other articles of artistic and utilitarian value from postharvest residual wood obtained from the local sawmill. Under a grant from the North American Fund for Environmental Cooperation, the SmartWood Program and its Mexican partner CCMSS are conducting a project in which forest product and market specialists will visit each of the certified ejidos to analyze their production capacity, market analysis strategies, and promotional and sales efforts and then provide direct assistance to the ejidos in securing domestic or export markets for their certified raw material or value-added products (CEC 1997).

Unión de Comunidades Forestales Zapoteco-Chinanteca (UZACHI)

Founded in 1989 as a union of indigenous communities (San Mateo Capulalpam de Mendez, Santiago Comaltepec, Santiago Xiacui, and La Trinidad Ixtlan), UZACHI was formed to protect community forest resources in the Sierra Juarez mountains of Oaxaca, Mexico. The UZACHI enterprise had its origins in a grassroots movement that began in the early 1980s to convince the government to drop a long-term concession to a pulp and paper company and transfer forest stewardship responsibility for twenty-one thousand hectares of upland pine-oak forest into the hands of local communities.

A government initiative in the 1960s to make Mexico independent in pulp and paper production led to environmentally destructive forest management dominated by the practice of high-grading, where only the largest and most valuable trees are cut. Old-growth pines were not protected, and large diameter sawtimber was not utilized for high-value products but instead was cut indiscriminately to supply an unprofitable pulp mill. Compounding its poor forest practices, the concession offered few benefits to the communities, due to the low wages and poor working conditions it offered.

The nonprofit organization Estudios Rurales y Asesoría Campesina (ERA) has provided technical assistance for improved forest development on the part of UZACHI and other communities in the Sierra Juarez region of Oaxaca (Chapela and Lara 1996). The UZACHI technical group developed and executed long-term forest management plans for each community, and ERA developed the comprehensive Geographic Information System, with a mapping and database component incorporating an in-depth ecological study, as well as management plans for timber extraction, transport, and milling. In 1996, SmartWood, working with its Mexican affiliate CCMSS, completed the assessment and certification of UZACHI.

The FSC-endorsed certification of UZACHI rests partially on the environmental benefits achieved through implementation of the communities' land management plan. Approximately one-third of the certified forest area is reserved for protection of watersheds, wildlife, recreation, and nursery seed stock. Timber management employed by the communities has helped to reverse the environmental damages caused by previous practices. For example, the UZACHI forest area has increased by five hundred hectares over the last two decades as a result of community decisions to rehabilitate abandoned agricultural sites and restrict new clearing of forest cover (Klooster 1997). Examples of environmental safeguards in timber extraction include design of logging roads to prevent damage to river beds, directional felling, and special attention to protecting seed trees and residual vegetation during cutting operations (Merino and Alatorre 1997). Higher costs and lower returns associated currently with UZACHI's low-impact timber management constitute investments in long-term improvements in the commercial productivity and ecological health of the forest.

The certification process has yielded a number of positive results for the UZACHI communities. Benefits have included enhanced public recognition within Mexico and internationally, improved resource inventories and mapping, and greater participation by community residents in forest management and conservation.

The prospect of increased access to European and North American markets was a significant factor behind UZACHI's taking steps to obtain certification. However, the communities have not yet developed the capacity to tap into international markets, and all UZACHI wood produced currently goes to the Mexican domestic market, principally to the communities themselves. The UZACHI communities harvest and mill several species of upland pine, including *Pinus psedostrobus, Pinus patula*, and *Pinus ayacahuite* for local markets. Part of their plan calls for product development and marketing of the region's abundant underutilized hardwoods such as various oaks (*Quercus laurina, Quercus crassifolia*), sweetgum (*Styraciflua* spp.), Pacific madrone (*Arbutus menziesii*), and avocado (*Persea* spp.). The communities have installed a kiln and a small workshop and are working with a furniture designer in Oaxaca to develop more value-added wood industries, including manufacturing of chairs and diskette boxes.

The UZACHI communities are understandably very interested in pursuing international market opportunities for FSC-endorsed wood, including furniture and other value-added products. In addition, the communities are exploring market openings for nontimber forest products such as gourmet mushrooms, medicinal plants, and pine resin. Ecotourism could also become increasingly important to the local economy because of the region's breathtaking scenery and biological richness; an ecotourism venture was recently launched in Itzlán de Juarez, a town neighboring the UZACHI communities in the Sierra Norte of Oaxaca. Nontimber forest products and ecotourism are viewed as potential sources of income derived from the sizable portion of the UZACHI-certified land area that is designated for environmental protection and is off limits to logging.

Creating Market and Policy Incentives for Improved Forest Management

Historically, market and policy signals have not favored the sort of environmentally sound, socially responsible forest management exemplified by the community forest areas operating with FSC endorsement in Mexico. On the contrary, the policy climate has for decades been biased in favor of agricultural crop and livestock production and against use of land for forestry (Carabias 1996). Current developments in Mexican law and policy, consumer markets for forest products, private investment, and international development financing may provide new opportunities for changing the incentive structure in ways that promote improved forest management.

Mexican Law and Policy

Over the last ten to fifteen years, trends in Mexican forest law and policy have been steadily oriented toward stimulating private investment in ejido and community forestry, as well as toward establishment of a plantation sector within Mexico's forest products industry. In 1986, Mexico enacted a forestry law that ended a thirty-year era of policy based on government-awarded concessions for timber extraction by large parastatal enterprises. The 1986 law represented a fundamental shift toward more direct community control over forest management (Chapela 1997).

In 1992, the government of President Carlos Salinas de Gortari completed sweeping changes to Article 27 of the Mexican Constitution. These reforms reversed agrarian policies dating from the Mexican Revolution that had guaranteed land redistribution to the country's rural sector. Among other features, the new version of Article 27 permits individual ejido members to sell their parcels of land and to enter into business partnerships with private investors (Barry 1995). The privatization of Mexico's ejidos was elaborated in the new Agrarian Law, enacted in 1992 to implement the Article 27 reforms. These major constitutional and legislative changes apply as much to the ejidos that are principally engaged in forest management as to those engaged in agricultural production.

A forestry law approved in 1992, together with regulations promulgated in 1994, largely removed the Mexican government's technical service role in forest management, permitting forest communities to obtain services from private sector specialists. The 1992 version of Mexico's *Ley Forestal* also promoted government incentives for timber plantations; these incentives were patterned on similar programs instituted in Chile (Silva 1997).

Plantation subsidies were a central focus of the forestry program carried out by the administration of President Ernesto Zedillo Ponce de León. Among other parts of the incentives package, investors in plantations are eligible for direct financial assistance from the government covering up to 65 percent of their costs incurred during the early years of their investment (SEMARNAP 1996). The government's basic objective is to stimulate domestic jobs and income by making Mexico self-sufficient in paper products and ultimately to transform the country into an international force in paper markets. Under the

Programa de Desarrollo de Plantaciones (PRODEPLAN), the government will administer a competitive enrollment period for each of seven years during the period 1997–2003, in which applicants will submit requests to be awarded funds from an annual budget for plantation subsidies totaling approximately $31 million.

Government and industry supporters of the plantations stimulus program note that Mexico has some eight million hectares of land, concentrated largely in the Gulf Coast region of southeastern Mexico, that would be well suited to timber plantations. Advocates of the program contend that plantations will promote environmental objectives by shifting logging pressures away from natural forests, restoring lands that have been degraded by unsustainable crop production or livestock grazing, enhancing wildlife habitat, and sequestering atmospheric carbon dioxide (Ignacio et al. 1997).

Whether timber plantations in Mexico yield significant conservation benefits will depend on local factors, such as whether the plantations are capable of providing biological links between areas of intact natural forest. Experience with extensive plantations in Brazil has shown the environmental impacts to be neither categorically good nor bad but rather dependent on site-specific circumstances (Viana 1997). Because they mainly produce fiber for pulp and paper markets, plantations would not directly substitute for logging in Mexico's natural forests, where commercial production is oriented more toward solid wood products (Paré 1997). Heavy subsidies for plantation investors, which dwarf the government's budgetary allocations for environmentally sound natural forest management, provide no incentive for Mexico's rural communities to maintain land in natural forest cover (Paré and Madrid 1997).

The plantations controversy figured prominently when forestry law reform re-emerged as one of the dominant environmental issues in Mexico during 1997. Following months of rancorous and highly partisan debate, the Mexican national legislature voted in April 1997 to approve changes proposed by the Zedillo Administration to the country's Ley Forestal. Opponents of the 1997 reforms charged that the government was putting the interests of transnational timber companies ahead of the long-term development needs of rural communities. Opponents complained also that the reforms would encourage unchecked planting of eucalyptus trees, which can deplete soils, reduce biological diversity, and require high chemical applications.

Leaving aside the political divisions, a closer look at the 1997 Ley Forestal reveals a number of significant provisions that reflect recommendations made by conservation and community forest advocates in Mexico. The law appears not so much to introduce new subsidies for timber plantations as to fortify social and environmental safeguards around the existing plantations subsidy and other timber stimulus programs the government had established prior to the 1997 legislative debate (Paré and Madrid 1997). In particular, the new law contains explicit language restricting the conversion of natural forests to monocultural timber plantations. The new law also imposes regulatory requirements designed to prevent illegal logging and illicit timber trade.

Markets for Certified Forest Products

Although certification shows great promise in marketing, existing FSC-endorsed community forest enterprises in Mexico have not yet seen dramatically enhanced market access as a result of their investment in becoming certified. For the moment, FSC-endorsed wood products from Mexico occupy a niche market focused on commercial or lesser known tropical hardwoods from the Yucatan Peninsula. Price premiums have been obtained within sales of certain certified products, but premiums have not been the main benefit of certification activities to date. Instead, forest certification in Mexico and other Latin American countries has been motivated principally by other considerations, including credibility, market penetration, and the hope of future price differentials (Bihun 1998).

In the near term, certified forest products from Mexico will continue to command only a negligible market share internationally. Nevertheless, the niche market penetration that Mexico's certified communities have achieved illustrates the potential for significant future growth.

In North America and Europe, there are currently more than two dozen chain-of-custody certified manufacturers and distributors dealing in certified wood products from Mexico. Most of these are small or medium-sized operations that carry tropical hardwoods and cater to emerging "green building" markets. Most of the distributors carry only a portion of their stock in certified goods. A small number of distributors or manufacturers carry some FSC-endorsed wood products from Mexico but have not yet obtained chain-of-custody certification.

The certified distributors of FSC-endorsed Mexican wood products fall into one or more of three categories: primary distributors, secondary distributors or manufacturers, and limited production or artisanal workshops. A leading company in the primary distributor category is the Berkeley, California–based EcoTimber International, whose business is based on trade in FSC-endorsed certified wood, as well as recycled or "rediscovered" wood products. With a warehouse facility of sixteen thousand square feet and annual sales exceeding $2 million, EcoTimber is the world's largest distributor of FSC-endorsed hardwoods.

Secondary distributors or manufacturers transform the lumber to finished products such as flooring, furniture, or shelving. Gibson Guitar Company of Nashville, Tennessee, and Martin Guitars of Nazareth, Pennsylvania, produce premium quality instruments that include wood from certified forests in Mexico. ABC Certified Forest Products of New Brunswick, Canada, carries lumber and furniture manufactured in Mexico from woods harvested from the FSC-endorsed forestry ejidos in Quintana Roo. In the past, International Hardwood Flooring, based in Philadelphia, has carried a line of certified flooring featuring several lesser known tropical hardwoods from Quintana Roo.

Limited production or artisanal workshops produce one-of-a kind studio or exhibition-quality furniture, architectural woodwork, or wooden accessories. Examples include the Massachusetts-based Karp Woodworks, which also carries FSC-endorsed lumber from Mexico, and Michael Elkan Studio, a box maker in Oregon.

The market for FSC-endorsed products from certified community forests in Mexico could get a significant boost from the international proliferation of "buyers groups" made up of companies committed to purchasing wood from independently certified, well-managed forests. The 1995 Plus Group, catalyzed by the World Wide Fund for Nature (WWF) in the United Kingdom, was the first concerted initiative to mobilize market demand for certified forest products. The group's eighty members account collectively for three billion pounds sterling in annual wood products trade and 15 percent of the wood consumption in the United Kingdom. Included in the group's membership are the United Kingdom's three largest chains of "do-it-yourself" home improvement stores, which together command a 30 percent market share (Sustainable Forestry Working Group 1998).

Companies that join the 1995 Plus Group are required, as a condition of membership, to phase in wood purchases from certified sources (Hansen 1997). Approximately thirty of the group's corporate members are currently purchasing certified wood and wood products, with five hundred certified product lines. Although only a modest fraction (roughly 3 percent) of the 1995 Plus Group's total wood trade currently consists of certified products, this small but growing amount is sending a powerful signal to the marketplace.

Other European buyers groups—in Belgium, Germany, the Netherlands, Spain, Switzerland, Latvia, and Austria—are being patterned along the lines of the 1995 Plus Group in the United Kingdom. Among the specific objectives of the newly developing Spanish buyers group known as Grupo 2000 is to facilitate greater trade with Mexico and Latin America to service the growing demands in Europe for certified forest products (Bihun 1998).

In 1997, the Certified Forest Products Council (CFPC) was launched as a nonprofit, voluntary business initiative dedicated to increasing the market for independently certified forest products in North America. The CFPC represents more than 160 companies, ranging from firms dealing in very large wood volumes such as Home Depot and Habitat for Humanity to smaller businesses such as Hard Wood Artisans of the Loft Bed Store and Colonial Craft. The CFPC membership also includes institutions, organizations, and individuals. As the organization grows, the CFPC could send a strong market signal in favor of independently certified forest management in Mexico. North American businesses facing growing demands for and tightening supplies of wood will be taking an increasingly hard look at independently certified sources in Mexico.

Private Investment in Forest Production

Notwithstanding the Mexican government's plantation subsidies and related incentives, private timber companies have yet to invest heavily in Mexico. The recent downturn in global paper markets appears largely to explain why large transnational firms such as International Paper and Champion International have not, at least for the moment, planned major expansions into Mexico (Feagans 1997).

There are signs that this situation could change dramatically in the future, however. During the government's first public offering of the plantation subsidies, held in the summer of 1997, more than one-third of the initial applications were submitted by private enterprises, with the remainder submitted predominantly by forestry ejidos and comunidades as well as small landowners (Cruz 1997). An estimated 40 percent of all applications sought government funds to support establishment of plantations of high-value native species such as cedar and mahogany in the tropical, southeastern part of Mexico (García 1997).

In addition, plans to invest in timber plantations in the Mexican states of Tabasco and Chiapas have been announced by Grupo Pulsar, which is a giant conglomerate, based in the northern Mexican city of Monterrey, that deals in a wide variety of products and services such as vegetables, seeds, baked goods, and insurance. Two U.S.–based firms, Simpson Corporation and Temple Inland, have signaled intentions to invest in plantation projects in southeastern Mexico (Feagans 1997).

The FSC principles and criteria may provide a promising basis for more successful social sector–private sector partnerships that resolve issues surrounding indigenous peoples' rights, unsustainable timber cutting, and other prevalent concerns within the management of Mexico's community forests and the establishment of plantations. Operating within the FSC framework, private investors could conceivably furnish badly needed capital for upgrades of milling and harvesting equipment, road improvements, and worker training, in tandem with independently verified compliance with forest management standards that protect community development and environmental interests. Such scenarios would help to fulfill objectives that have been stated repeatedly in Mexican forestry laws and policies but have never been achieved in practice.

International Financing

Well-targeted international financing can help make it possible for Mexican forest communities to improve management practices and to obtain certification. To date, a number of bilateral foreign assistance agencies, as well as private foundations, have provided significant financial support for community forestry in Mexico. Examples include the German Development Assistance Agency, the United Kingdom's Overseas Development Administration, the Inter-American Foundation, the John D. and Catherine T. MacArthur Foundation, the Rockefeller Foundation, the Ford Foundation, and the Moriah Fund. There is potential for various international institutions to provide new or increased support to cover communities' costs such as those associated with land management planning, equipment upgrades, product development, quality control, marketing, tracking the "chain of custody" for certified products, training, and fees for certification evaluations and monitoring. The following institutions could play important roles.

World Bank

Opportunities may exist to build upon an existing World Bank project to promote sustainable community forestry in the state of Oaxaca. Officially announced in 1997, this $50 million project is designed principally to support forestry communities and ejidos through technical assistance for land management planning and related activities, training of private sector forestry technicians in environmentally sound management practices, and commercial development and promotion of nontimber forest products (World Bank 1995b). The intent is to replicate positive results achieved in Oaxacan forest communities in other Mexican states with important forest resources. Additional opportunities for Mexico may reside in a recently announced collaboration between the World Bank and WWF, which, among its leading objectives, seeks to achieve independently certified management of two hundred million hectares of forest land globally by the year 2005.

Global Environment Facility

A potentially important source of funding for FSC-endorsed community forestry in Mexico is the Global Environment Facility (GEF), which is the world's only multilateral dedicated fund for the environment, providing grants to developing countries to protect global resources. Independently certified community forest management in Mexico fits well within the GEF's operational strategy for biological diversity. Among other priorities, the Facility's strategy document refers specifically to "development of sustainable use methods in forestry"; creation of "participatory schemes for natural resource management . . . by local communities, indigenous groups, and other sectors of society"; and "conservation and development projects in which social needs are integrated into project design" (GEF 1996).

Commission for Environmental Cooperation

Another important potential source of increased international financing for community forestry in Mexico is the Commission for Environmental Cooperation (CEC), which was established in 1993 as part of the environmental provisions surrounding the North American Free Trade Agreement. One of the CEC's accomplishments has been creation of a North American Fund for Environmental Cooperation (NAFEC) to support sustainable development projects in local communities throughout the three participating countries. Several projects approved in 1997 for NAFEC funding relate directly to FSC-endorsed forest management and products in Mexico (CEC 1997).

Inter-American Development Bank

Funding opportunities for FSC-endorsed community forestry in Mexico may exist within the Inter-American Development Bank (IDB), which has become the largest source of development financing in Latin America. A recent document from the IDB's environment division mentions "support for identification of appropriate technologies in natural forest management" among the Bank's priorities regarding management of biological resources (IDB 1995).

Conclusion

Enormous challenges, as well as significant potential opportunities, confront Mexico's forests and forest residents. One part of a comprehensive solution resides in community-based forest management that provides lasting benefits to rural development and environmental quality. Environmentally sound community forestry, as well as the emerging tool of FSC-endorsed certification, deserves strong support from governments, international institutions, private sector businesses, and NGOs.

Acknowledgments   Research for this paper has been made possible, in part, by generous grants from the Moriah Fund, the Rockefeller Brothers Fund, and the Wallace Global Fund. The support of the 400,000 members of the Natural Resources Defense Council (NRDC) was also invaluable to the completion of this project. The authors gratefully acknowledge the helpful advice and information supplied by Enrique Alatorre, Alfonso Argüelles, Kim Batchelder, David Bray, Francisco Chapela, Richard Donovan, Emily Goldman, Leticia Merino, Mike Kiernan, Daniel Klooster, Sergio Madrid, Laura Snook, and Pete Taylor. The authors are solely responsible for the contents of this essay and would welcome comments and questions to help inform subsequent editions of this chapter or related activities.

REFERENCES

Arguelles, A. 1993. "Conservación y Manejo de Selvas en el Estado de Quintana Roo, México." Unpublished report of the Acuerdo México-Alemania, Chetumal, Quintana Roo, Mexico.

Barry, T. 1995. *Zapata's Revenge: Free Trade and the Farm Crisis in Mexico*. South End Press, Boston, Mass.

Beyer, K., SmartWood Program. 1998. Personal communication.

Bihun, Y. A. 1998. Good climate for certification. *Timber Trades Journal* 383:14–16.

Bowles, I., and G. Prickett. 1994. *Reframing the Green Window*. Conservation International and Natural Resources Defense Council, Washington, D.C.

Bray, D. 1997. La Reconstrucción Permanente de la Naturaleza. In *Semillas Para el Cambio en el Campo: Medio Ambiente, Mercados y Organización Campesina*. Mexico City. Pp. 3–17.

Bryant, D., et al. 1997. *The Last Frontier Forests: Ecosystems and Economies on the Edge*. World Resources Institute, Washington, D.C.

CEC. 1997. *EcoRegion*. Secretariat Bulletin 6. Commission for Environmental Cooperation, Montreal QC, Canada.

Carabias, J. 1996. Address at the Opening of the Forest Stewardship Council's First General Assembly, Oaxaca, Mexico.

Chapela, F., and Y. Lara. 1996. *La Planeación Comunitaria del manejo del Territorio*. and, Oaxaca, Mexico.

Chapela, G. 1997. El cambio liberal del sector forestal en México. In *Semillas Para el Cambio en el Campo: Medio Ambiente, Mercados y Organización Campesina*. Mexico City, Mexico. Pp. 37–56.

Clifford, F. 1997. Which comes first—food or the forest? *Los Angeles Times*, July 24, p. 1.

Cruz, A. 1997. Concluyó la subasta para establecer plantaciones forestales comerciales. *La Jornada*, June 28, p. 46.

FAO. 1997. *State of the World's Forests: 1997*. United Nations Food and Agriculture Organization, Rome, Italy.

Feagans, B. 1997. Forest slump: Laws that opened Mexico for lucrative tree plantations have yet to blossom into a bloom. *U.S.–Mexico Business.* Summer, pp.

Galletti, H. 1998. The Maya Forest of Quintana Roo: Thirteen Years of Conservation and Community Development. In *Timber, Tourists and Temples: Conservation and Development in the Maya Forest of Belize, Guatemala, and Mexico,* edited by R. Primack et al. Island Press, Washington, D.C. Pp. 33–46.

GEF. 1996. *Operational Strategy.* Global Environment Facility, Washington, D.C.

García, M. 1997. Grupos ejidales, comunales y parvifundistas buscan participar en la licitación de plantaciones comerciales. *La Jornada,* May 20, p. 40.

Hansen, E. 1997. Forest certification and its role in marketing strategy. *Forest Products Journal* 47:16–22.

IDB. 1995. *The IDB's New Orientation towards the Environment: Objectives and Functions of the Environment Division.* Inter-American Development Bank, Washington, D.C.

Ignacio, E., et al. 1997. Plantaciones forestales en México: Una alternativa para el manejo sustentable de los recursos naturales. In *Bosques y Plantaciones Forestales.* Cuadernos Agrarios, Mexico City, Mexico. Pp. 34–40.

Kiernan, M., and C. Freese. 1997. Mexico's Plan Piloto Forestal: The search for balance between socioeconomic and ecological sustainability. In *Harvesting Wild Species: Implications for Biodiversity,* edited by C. Freese. Johns Hopkins University Press, Baltimore.

Klooster, D. 1997. Conflict in the commons: Commercial forestry and conservation in Mexican indigenous communities. Ph.D. dissertation, University of California, Los Angeles.

Merino, L. 1997. Organización Social de la Producción Forestal Comunitaria. In *Semillas Para el Cambio en el Campo: Medio Ambiente, Mercados y Organización Campesina.* Mexico City, Mexico. Pp. 141–154.

Merino, L., and G. Alatorre. 1997. Hacia la sustentabilidad del manejo de los bosques templados: Capulalpam, Oaxaca y el ejido de Rosario de Xico, Veracruz. In *El Manejo Forestal Comunitario en México y sus Perspectivas de Sustentabilidad,* edited by L. Merino. Cuernavaca, Mexico. Pp. 109–129. National Autonomous University of Mexico.

Paré, L. 1997. Las plantaciones forestales de eucalipto en el sureste de México: ¿Una prioridad nacional? In *Bosques y Plantaciones Forestales.* Cuadernos Agrarios, Mexico City, Mexico. pp. 41–62.

Paré, L., and S. Madrid. 1997. Las modificaciones a la ley forestal: ¿Solamente apoyos y subsidies a plantadores forestales transnacionales? In *Bosques y Plantaciones Forestales.* Cuadernos Agrarios, Mexico City, Mexico. Pp. 11–15.

Perry, J. 1991. *The Pines of Mexico and Central America.* Timber Press, Portland, Oregon.

SEMARNAP. 1996. *Síntesis Ejecutiva: Programa Forestal y de Suelo 1995–2000.*

Silva, E. 1994. Thinking politically about sustainable development in the tropical forests of Latin America. *Development and Change* 25:697–721.

———. 1997. The politics of sustainable development: Native forest policy in Chile, Venezuela, Costa Rica and Mexico. *Journal of Latin American Studies* 29:457–493.

Snook, L. 1997. Implicaciones de la tenencia comunitaria y los recientes cambios en las políticas. In *Semillas Para el Cambio en el Campo: Medio Ambiente, Mercados y Organización Campesina.* Mexico City, Mexico. Pp. 19–35.

———. 1998. Sustaining harvests of mahogany (*Swetenia macrophylla King*) from Mexico's Yucatan forests: Past, present and future. In *Timber, Tourists and Temples: Conservation and Development in the Maya Forest of Belize, Guatemala, and Mexico*, edited by R. Primack et al. Island Press, Washington, D.C.

Sustainable Forestry Working Group. 1998. *Sustaining Profits and Forests: The Business of Sustainable Forestry.* John D. and Catherine T. MacArthur Foundation, Chicago, Illinois.

Upton, S., and S. Bass. 1995. *The Forest Certification Handbook.* Earthscan Publications, London, England.

Viana, V. 1997. Plantaciones industriales en Brasil: Lecciones para el desarrollo sostenible. In *Bosques y Plantaciones Forestales.* Cuadernos Agrarios, Mexico City, Mexico. Pp. 99–104.

Whigham, D., et al. 1998. Dynamics and ecology of natural and managed forests in Quintana Roo, Mexico. In *Timber, Tourists, and Temples: Conservation and Development in the Maya Forest of Belize, Guatemala, and Mexico*, edited by R. Primack. et al. Island Press, Washington, D.C. Pp. 267–81.

World Bank. 1995a. *Mexico: Resource Conservation and Forest Sector Review Report no. 13114-ME.* Natural Resources and Rural Poverty Operations Division, World Bank, Washington, D.C.

———. 1995b. *Project Information Document: Mexico Community Forestry.* World Bank, Washington, D.C.

World Forest Institute. 1994. *Mexico and the Wood Products Industry.* World Forestry Center, Portland, Oregon.

Zabin, C., and P. Taylor. 1997. *Quintana Roo Forestry Management Project Consultants' Report Summary.* Overseas Development Administration.

10 Options for Conserving Biodiversity in the Context of Logging in Tropical Forests

Richard Rice
Cheri Sugal
Peter C. Frumhoff
Elizabeth Losos
Raymond Gullison

Current patterns of tropical timber production pose one of the most significant threats to biodiversity of all natural resource industries. Logging, by its very nature, can seriously affect the structure and species composition of forests (Bawa and Seidler 1998). Moreover, the area of tropical forest affected by logging is vast: A. G. Johns (1997) estimates that 31 percent of remaining tropical forest is officially allocated to timber production, considerably more than the roughly 9 percent that is at least nominally under stricter protection. The area of production forest is also growing, particularly in the tropical Americas, with increased foreign investment in forestry (Bowles et al. 1998) and expanding domestic and international demand for tropical hardwood products (FAO 1993; Uhl et al. 1997).

Preceding chapters have focused on the overall threat of logging to tropical biodiversity, the role of community management and timber certification in conservation, and a case study of the profitable development of plantations to meet wood and pulp supplies. In this chapter, we address the basic question of how best to achieve concrete conservation results in the context of logging in natural tropical forests.

In recent years, most efforts to promote conservation in tropical forests have focused on one of two approaches—outright protection of high-priority areas and natural, or "sustainable," forest management (NFM). Outright protection, through parks and reserves, is an essential element in biodiversity conservation. Strict protection is not always politically or economically feasible, however. Areas that have valuable stocks of commercial timber, for example, will be under pressure for logging. In other areas, pressure on forested lands for conversion to agriculture is high (van Schaik et al. 1997; WRI et al. 1998).

Outside protected areas, the currently favored approach to protecting tropical forests is NFM. NFM typically combines harvesting guidelines designed to increase the growth of commercial tree species with reduced-impact logging techniques aimed at lowering the ecological impacts of timber harvests. In principle, NFM should contribute to conservation by making logging less destructive and by stabilizing wood production in a given area, thereby maintaining forest cover and reducing pressure on other forests (Buschbacher 1990). Our belief is that much of the conventional wisdom on the benefits of this approach to conservation is misplaced and that more direct approaches—protecting areas of high biodiversity value before logging where possible, and retiring lightly logged timber concessions—will ultimately prove more effective.

Identifying Priority Areas for Protection

Tropical forests vary widely in their potential contribution to biodiversity conservation. For this reason, an assessment of the biological distinctiveness of a particular forest is an important step in the selection of an appropriate conservation strategy. Some forests contain a far more distinctive biota than others. They may possess, for example, substantially greater levels of species richness (the number of species per unit of area), a greater degree of endemism (a measure of the extent to which species exist only within a locally circumscribed area), or higher levels of beta diversity (a measure of the rate of change in species composition across the forest landscape) (Frumhoff and Losos 1998).

One approach to assessing the value of different forests for biodiversity is provided by E. Dinerstein and colleagues (1995), who compared the biological distinctiveness of tropical broadleaf forests (and other types of natural habitat) on an "ecoregional" scale in Latin America. They concluded that, of the fifty-five ecoregions evaluated, sixteen contained "globally outstanding" biological diversity, thirteen "regionally outstanding," eighteen "bioregionally outstanding," and eight "locally important." The Napo moist forest running along the western arc of the Amazon is an example of a globally outstanding ecoregion; surveys of several different taxa provide evidence that it contains one of the world's most distinctive biotas (Prance 1977).

Another approach to establishing conservation priorities focuses on biodiversity "hotspots." Conceived by Norman Myers (1988, 1990) and subsequently refined by CI (Mittermeier et al. 1998), the hotspot

approach uses species richness, endemism—particularly of plants—
and degree of threat as its main criteria. The most recent global hotspot analysis recognizes twenty-five priority regions for biodiversity
conservation, seventeen of which include tropical forests. These areas
occupy less than 2 percent of the world's land surface but contain 46
percent of the diversity represented by endemic plants and more than
75 percent of the terrestrial animals listed by the IUCN as globally
threatened (Mittermeier et al. 1998).

Broad-scale assessments, such as ecoregional and hotspot analyses,
provide an essential first step in the process of identifying timber production forests of the highest value for biodiversity conservation. Ideally, these areas should be protected not only from deforestation and
fragmentation but also from the direct impacts of logging and silvicultural treatments on forest structure and species composition. Accordingly, these forests should be considered high-priority candidates for
full protection or lower impact (primarily nontimber), multiple use
management. Lower impact management could include activities
such as nontimber forest product extraction, ecotourism, biodiversity
prospecting, and carbon sequestration (Frumhoff and Losos 1998).

As appropriate to the size of the concession and consistent with
conservation needs, whole concessions might be reallocated in this
manner to less destructive uses. Alternatively, priority habitat for
conservation might be designated for protection within a larger concession (Hardner and Rice 1999). Such "conservation set-asides" are
already being developed in some countries. In Bolivia, for example,
timber producers can now receive tax benefits for protecting up to 30
percent of a concession (Gaceta Oficial de Bolivia 1996). Since these
areas are typically selected for their low commercial value, however,
they may not necessarily contain forest of high value for biodiversity.
To better ensure that set-asides make an important contribution to
conservation, their size and locations should therefore be based on
independent, site-specific assessments.

Although desirable, protection of forests prior to logging is not always practical, particularly if the forest has economically attractive
concentrations of valuable commercial species. In these situations,
the opportunity cost of conservation—or the earnings that landowners would need to forgo in order to ensure forest protection—is high.
Most of these areas are also characterized by limited government capacity for enforcement of harvest restrictions. We discuss the viability
of promoting NFM under these conditions hereafter.

Investing in Natural Forest Management

Outside protected areas, the current standard approach to achieving
biodiversity conservation in tropical forests is NFM. One of the over-

arching conservation goals of promoting NFM is to reduce the rate of deforestation by showing that the long-term production of timber is an economically viable alternative to clearing forests for crops and livestock (Johnson and Cabarle 1993; Dickinson et al. 1996). Though popular among conservationists and the international donor community, this approach has not proven very effective in practice. NFM has yet to be adopted as a viable alternative to conventional forms of logging, let alone serving as an effective buffer against more destructive uses of the land. Despite years of effort and hundreds of millions of dollars in public and private investments, almost no logging in the tropics outside plantations can be considered sustainable.

At present, most logging companies in the tropics engage in the rapid harvest of a limited number of valuable tree species because it is profitable, often far more than the delayed harvest and investment in commercial regeneration needed for long-term timber production. This is because the return from investments in future timber production—a function largely of the growth rate of commercial species, the rate of change in timber prices, and the discount rate—is commonly lower than that earned by rapidly harvesting marketable trees and investing the profits elsewhere (Kishor and Constantino 1993; Vincent 1995; Rice et al. 1997; Reid and Rice 1997). Moreover, government administrative capacity and political will to limit harvests through enforcement of cutting restrictions is often extremely limited (Hardner and Rice 1999).

Where these conditions prevail, it will be very difficult to motivate timber producers to stay in the forest for the long term. That is, where financial incentives and limited enforcement capacity favor rapid logging with little or no investment in commercial regeneration, it will most likely continue despite the best efforts of internationally funded programs to provide technical assistance and training in NFM. Under these circumstances, limited conservation dollars would be better invested in other strategies or in other forests where conditions are more favorable to long-term timber production.

In addition to the financial and institutional obstacles just noted, there are limits to how effective any type of logging (no matter how well regulated) can be in promoting the conservation of biological diversity. Indeed, the impacts of logging on forest biodiversity are often substantial (Frumhoff 1995; Bawa and Seidler 1998). Long-term forest management typically results in a simplification of both the structure and composition of forests (e.g., a reduction in the vertical "layering" of vegetation and in the number of trees and other species present). The extent of this simplification is affected by the harvest intensity or the number of trees cut per hectare, by the harvest frequency, and by the type and extent of postharvest silvicultural "treatments," cuttings designed to open the forest canopy to provide more light and nutrients to targeted species.

Intensive timber harvesting changes the structure of unlogged forests by causing large areas to revert to secondary succession, exchanging the structurally complex canopies typical of many unlogged forests for the simplified structure and greater openness characteristic of a younger, regenerating forest (Whitmore 1988). In a commercial setting such as NFM, this loss of forest complexity is essentially permanent since the interval between commercial timber harvests is usually far less than the time required for primary forest characteristics to re-emerge (Horne and Hickey 1991).

Ironically, in some cases, the structural and compositional changes caused by NFM can be far more extensive than those caused by conventional logging due to the harvest of a wider variety of commercial tree species and the need for postharvest treatments to promote growth and regeneration in the residual stand (Howard et al. 1996; Rice et al. 1997). For this reason, management interventions that focus more on reducing the ecological impacts of logging than on achieving successful regeneration of commercial tree species may under certain conditions be preferable from a conservation perspective (Frumhoff and Losos 1998). Such "reduced-impact" logging techniques may include careful planning of skid trails to reduce distance traveled and to minimize loss of forest cover and soil erosion, and directional felling of harvested trees to minimize damage to the surrounding forest (Johns 1997). The profitability of some conventional logging operations can be increased by introducing reduced impact logging practices (Barreto et al. forthcoming). Under these conditions, such measures might be undertaken voluntarily, with correspondingly lower investments required in control and enforcement.

As a mechanism for promoting conservation, however, these techniques are not without their limitations. First of all, reduced-impact logging interventions typically do not involve restricting the overall volume of timber that is harvested (de Graaf and Poels 1990; Putz and Pinard 1993). Where harvest levels are high, logging that is "reduced" in impact might therefore still result in damage that is unacceptably high from the point of view of conservation. In addition, even when harvest levels are more modest, continued logging (with or without the application of these techniques) will unavoidably produce a progressive simplification of the forest structure (Barreto et al. forthcoming). Finally, without the direct silvicultural investments required to ensure regeneration under NFM (including prolonging the rotation period), reduced-impact logging might not be sustainable in the long term. As noted earlier, including these costly silvicultural investments makes NFM financially unattractive and difficult to enforce.

Achieving More "Bang for the Buck"

Given the substantial constraints to utilizing NFM as a conservation strategy, governments, conservationists, and donor organizations should consider alternative ways to achieve their conservation goals. One alternative that has received surprisingly little attention is to specifically consider protecting some forests that have already been logged. The advantage of this approach from an economic perspective is that the "opportunity cost" of protection can be dramatically lower than for uncut forests (which, in turn, should translate into stronger political support). Ecologically, this approach is superior to continued logging because the process of forest simplification is halted and forest recovery is initiated.

The opportunities presented by this approach for tropical forests derive in part from their extremely high diversity. Unlike forests in the temperate zone, most tropical forests are quite heterogeneous and often have only a limited number of marketable species. This number is further reduced where the local demand for timber is small and transport costs are high, as is the case in many areas of the Amazon Basin. Under these conditions, market forces dictate that logging is quite selective, often much more so than the harvests required by NFM. Instituting protection after logging in these cases takes advantage of the fact that the bulk of the timber value may be tied to only a small percentage of the trees.

Even where many species are cut and the damage from logging is significant, however, protection may still lead to greater conservation benefits than continued logging. This is because a cessation of logging will allow the conservation value of the forest to grow over time as it regains its former structural and compositional characteristics (Horne and Hickey 1991).

One example of the implementation of this approach is in the southern Amazon Basin, where three electric utility companies in partnership with a U.S. conservation organization financed the conversion of roughly 630,000 hectares of timber concessions into an extension of Bolivia's Noel Kempff Mercado National Park (Shirley 1997). Full protection, in this instance, was considered a priority, despite the fact that the area had been previously logged, because past logging had been highly selective and the area is adjacent to an existing national park. The cost of acquiring the land was also extremely low (roughly $2.50 per hectare) since most of the area's high value tree species had already been removed (Petterson 1998).

An explicit assessment of the benefits of projects such as the Noel Kempff extension is provided by J. B. Cannon and his colleagues (1998), who used a computer simulation of the Chimanes Forest in Bolivia to compare protection of a forest after logging with other

Table 10.1 Ranking Conservation and Development Options by Selected Criteria, Chimanes Forest, Bolivia

Option	Structure Retained	Composition Retained	Profit[1]	Opportunity Cost[1]
	(relative to unlogged forest)		(relative to conventional logging)	
Protection of unlogged forest	100%	100%	0%	100%
Conventional logging, then protection[2]	95%	99%[3]	91%	9%
Conventional logging, no protection[4]	62%[5]	98%[6]	100%	0%
NFM[4]	33%[5]	80%[6]	37%	63%

1. Assuming a discount rate of 15 percent.
2. Protection after five years of logging.
3. In year five (average of years 1 to 10).
4. After fifty years of logging.
5. Average of years 41 to 50.
6. In year 50.

available options (specifically, outright protection, conventional profit-maximizing logging, and NFM). Profit was used in the study to measure economic performance, and structural and compositional changes (relative to unlogged forest) were used to measure conservation benefits. The study's findings were as follows.

In comparing structural changes, Cannon and his colleagues found that conventional logging actually caused less damage than NFM because the density of the primary commercial tree species in this area is extremely low: roughly one to two trees every five hectares. Rather than harvesting a wide variety of species on every hectare—as in the NFM option—conventional logging captured far more value with fewer trees by concentrating on only a small number of valuable species.

The findings of the simulation were similar when the compositional changes associated with conventional practices were compared with those caused by NFM. The reason relates to the fact that, as in many tropical forests, noncommercial tree species in the Chimanes Forest comprise the bulk of the tree diversity. In this case, seventy-eight of the forest's ninety-six species have no commercial potential (Howard et al. 1996). NFM systematically favored the relatively small number of commercial species at the expense of noncommercial species, reducing the forest's tree diversity by 20 percent. Conventional logging, in contrast, depressed the populations of only a few high-value species, causing little overall change in forest composition (Cannon et al. 1998).

Of the various options involving timber production (i.e., all of the options but outright protection), the least damage was caused by protection following a brief period of conventional logging. Ironically, this option was also one of the most profitable. With protection following logging, 95 percent of the structure of the unlogged forest was retained at a cost of only 9 percent of the maximum profits available from conventional logging. NFM, on the other hand, maintained only a third of the structure of the unlogged forest, at a cost of nearly two-thirds of the profits available from conventional logging (Cannon et al. 1998).

These figures suggest that for the Chimanes Forest, protection after a short period of logging may be more than twenty times as cost-effective as NFM at achieving conservation objectives in terms of forgone profits. In other words, for a given reduction in profit, twenty times as much forest area could be protected after logging as could be placed under NFM. Alternatively, for a given area placed under NFM, an equal area could be protected following logging for one-twentieth the reduction in profits.

Analyses such as Cannon's (1998) have not yet been performed at sites where forests differ substantially from Chimanes. Dipterocarp forests in Malaysia and Indonesia, for instance, tend to have a much higher density of commercial trees, which leads to more damaging conventional logging. In other tropical production forests, especially those in Central Africa, foreign logging companies have moved away from selective logging in favor of higher-intensity harvests. More research is therefore needed to determine how general the findings for the Chimanes Forest are. Our expectation, however, is that while the absolute levels of profit and damage will vary according to site-specific conditions, the ranking of the various options will probably remain the same across a wide range of circumstances in Amazonia and perhaps even throughout the tropics. The available evidence, for example, shows that conventional logging is consistently more profitable than NFM despite site-specific changes in the proportion of high-value tree species (Cannon et al. 1998). Similarly, even where conventional logging is more intensive than NFM, and therefore does more damage during an initial harvest, protection following a brief period of logging, may still yield greater conservation benefits than NFM over the long run due to subsequent forest recovery.

Broadening the criteria by which different options are judged is only likely to strengthen these conclusions. Adding carbon storage value, for example, would also favor selective logging followed by protection if timber harvests are not displaced elsewhere. Carbon storage depends largely on the volume of wood retained in the forest. With protection following selective logging, more of the structure and composition—and therefore more of the carbon—is retained. Similarly, continued logging may increase the risk and extent of forest fires, which can release significant amounts of carbon into the atmosphere and cause additional loss of biodiversity.

Conclusion

In recent years, conservation and development organizations have focused their attention on outright forest protection for areas of high value for biodiversity. Elsewhere, NFM has been promoted as a way to reap both economic and conservation benefits from the forest over the long term. In theory, NFM reconciles traditional economic activity with protection of ecosystems and would appear to be an ideal solution to the economic constraints to setting aside large tracts of land.

In practice, however, NFM can cause significant changes to the structure and composition of forests and is rarely the most profitable approach to logging. Reduced-impact logging may eliminate some of these shortcomings but has its own limitations as well, including the fact that over the long term it is not necessarily sustainable and will inevitably cause a progressive simplification of the forest.

One alternative that has received little attention to date is to focus some protection investments specifically on areas that have already been logged. This approach will yield the greatest conservation benefits at the lowest cost when the majority of the timber value is tied to a small percentage of the trees—a situation that characterizes millions of hectares of forest throughout the tropics. When compared with NFM, protection following a short period of logging can yield both significantly greater economic returns and less environmental damage. In an era of shrinking budgets and rapidly expanding logging activity, the cost-effectiveness of this and other alternatives to more traditional approaches merits careful consideration.

REFERENCES

Banco Central de Bolivia, Gerencia de Estudios Economicos. 1994. *Boletin Estadistico* no. 283, September (for January and February 1994). Cuadro no. 5,01A. Tasas Activas de Interes Anual en el Sistema Bancario Nacional (Moneda Extranjera ME, Efectivas). La Paz. P. 61.

———. 1995a. *Boletin Estadistico* no. 285, March (for March–December 1994 and January–September 1995). Cuadro no. 5,01B. Tasas de Interes Activas Negociadas en el Sistema Bancario Nacional (Moneda Extranjera, Efectiva). La Paz. P. 63.

———. 1995b. *Informacion Estadistica Mensual* no. 1.726, December 12. Cuadro no. 1.7.4, P. 30. Tasas de Interes Referenciales, col. 9 (Activa Referencial, Dolares). Quito. P. 30.

Banco Central de Reserva del Peru, Gerencia de Estudios Economicos. 1996. *Nota Seminal* no. 3, January 19. Cuadro 29. Tasas de

Interes Activas y Pasivas Promedio de la Banca Multiple (Moneda Extranjera, Prestamos Activas, mas de 360), Lima, Peru.

Bawa, K., and R. Seidler. 1988. Natural forest management and the conservation of biological diversity in tropical forests. *Conservation Biology* 12:46–55. February.

Barreto, P., P. Amaral, E. Vidal, and C. Uhl. Forthcoming. Costs and benefits of forest management for timber production in eastern Amazon.

Bowles, I., A. Rosenfeld, C. Sugal, and R. Mittermeier. 1998. *Natural Resource Extraction in the Latin American Tropics: A Recent Wave of Investment Poses New Challenges for Biodiversity Conservation*. Policy brief no. 1. Conservation International, Washington, D.C.

Boxman, O., N. R. de Graaf, J. Hendrison, W. B. Jonkers, R. L. H. Poels, P. Schmidt, and R. Tjon Lim Sang. 1985. Towards sustained timber production from tropical rain forests in Suriname. *Netherlands Journal of Agricultural Science* 33:125–132.

Brown, P., B. Cabarle, and R. Livernash. 1997. *Carbon Counts: Estimating Climate Change Mitigation in Forestry Projects*. World Resources Institute, Washington, D.C.

Buschbacher, R. J. 1990. Natural forest management in the humid tropics: Ecological, social, and economical considerations. *Ambio* 19:253–258.

Cannon, J. B., R. E. Gullison, and R. E. Rice. 1998. *Evaluating Strategies for Biodiversity Conservation in Tropical Forests*. Development Research Group, World Bank, Washington, D.C.

Council of Economic Advisors (CEA). 1996. *Economic Report to the President*. U.S. Government Printing Office, Washington, D.C.

De Graaf, N. R., and R. L. H. Poels. 1990. The celos management system: A polycyclic method for sustained timber production in South American rain forest. In *Alternatives to Deforestation: Steps Towards Sustainable Use of the Amazon Rain Forest*, edited by A. B. Anderson. Columbia University Press, New York. Pp. 116–127.

Dickinson, M. B., J. C. Dickinson, and F. E. Putz. 1996. Natural forest management as a conservation tool in the tropics: Divergent views on possibilities and alternatives. *Commonwealth Forestry Review* 75:309–315.

Dinerstein, E., D. M. Olson, D. J. Graham, A. L. Webster, S. A. Primm, M. P. Bookbinder, and G. Ledec. 1995. *A Conservation Assessment of the Terrestrial Ecoregions of Latin America and the Caribbean*. World Wildlife Fund and World Bank, Washington, D.C.

Editora BBT Ltda. 1996. *Cenarios: Analise e Projecao Economica*. Ano 8, no. 72 (January). Custo Efetivo do Emprestimos. P. 6.

FAO. 1993. *Forest Resources Assessment 1990: Tropical Countries*. United Nations. Food and Agriculture Organization, Rome.

———. 1997. *State of the World's Forests 1997*. United Nations. Food and Agriculture Organization, Rome.

Forest Stewardship Council, U.K. Working Group (FSC). 1998. *Forests Certified by FSC-Accredited Certification Bodies*. FSC Website: www.fsc-uk.demon.co.uk/forest.html. Updated January 28.

Frumhoff, P. C. 1995. Conserving wildlife in tropical forests managed for timber. *BioScience* 45:456–464.

Frumhoff, P. C., and E. Losos. 1998. *Setting Priorities for Conserving*

Biological Diversity in Tropical Timber Production Forests. Union of Concerned Scientists, Cambridge, Mass., and Smithsonian Center for Tropical Forest Science, Washington, D.C.

Gaceta Oficial de Bolivia. 1996. Ley forestal no. 1700, articulo 29, deg. 3f). Ano 36, no. 1944 (July 12). La Paz, Bolivia.

Gerwing, J. J., J. S. Johns, and E. Vidal. 1996. Reducing waste during logging and log processing: Forest conservation in eastern Amazon. *Unasylva* 187(47):17–25.

Gullison, R. E., S. N. Panfil, J. J. Strouse, and S. P. Hubbell. 1996. Ecology and management of mahogany (*Swietenia macrophylla* King) in the Chimanes Forest, Beni, Bolivia. *Botanical Journal of the Linnean Society* 122:9–34. September.

Hardner, J. J., and R. Rice. 1999. *Rethinking Forest Concession Policy in Latin America.* Inter-American Development Bank, Washington, D.C.

Holloway, J. D., A. H. Kirk Spriggs, and C. V. Khen. 1992. The response of some rain forest insect groups to logging and conversion to plantation. *Philosophical Transactions, Royal Society of London* B335:425–436.

Horne, and Hickey. 1991. Review: Ecological sensitivity of Australian rainforests to selective logging. *Australian Journal of Ecology* 16:119–129.

Howard, A. F., R. E. Rice, and R. E. Gullison. 1996. Simulated economic returns and environmental impacts from four alternative silvicultural prescriptions applied in the Neotropics: A case study of the Chimanes Forest, Bolivia. *Forest Ecology and Management* 89:43–57.

International Monetary Fund (IMF). 1995. *International Financial Statistics.* IMF, Washington, D.C.

International Tropical Timber Organization (ITTO). 1996. *Annual Review and Assessment of the World Tropical Timber Situation.* ITTO,Yokohama, Japan.

Johns, A. G. 1997. *Timber Production and Biodiversity Conservation in Tropical Rain Forests.* Cambridge University Press, Cambridge, England.

Johnson, N., and B. Cabarle. 1993. *Surviving the Cut: Natural Forest Management in the Humid Tropics.* World Resources Institute, Washington, D.C.

Kishor, N. M., and L. F. Constantino. 1993. *Forest Management and Competing Land Uses: An Economic Analysis for Costa Rica.* LATEN Dissemination note 7. World Bank, Washington, D.C.

Kollert, W., K. Uebelhor, and M. Kleine. 1995. *Financial Analysis of Natural Forest Management on a Sustained-yield Basis: A Case Study for Deramakot Forest Reserve.* Report 200.

Mason, D. 1996. Responses of Venezuelan understory birds to selective logging, enrichment strips, and vine cutting. *Biotropica* 28(3): 296–309.

Myers, Norman. 1988. Threatened biotas: Hotspots in tropical forests. *Environmentalist* 8:187–208.

———. 1990. The biodiversity challenge: Expanded hotspots analysis. *Environmentalist* 10:243–256.

Mittermeier, R. A., N. Meyers, J. B. Thomsen, G. A. B. da Fonseca, and S. Olivieri. 1998. Biodiversity hotspots and major tropical wilderness areas: Approaches to setting conservation priorities. *Conservation Biology* 12(3).

Muller, E., J, president, De Surinaamsche Bank, nv. 1995. Personal communication, December.

Petterson, J. 1998. "The Nature Conservancy Bolivia and Partners Sign Agreement on Climate Change Project: Corporations Invest in Forest Protection, Potential Carbon Emission Credits." Unpublished fact sheet. March 3.

Prance, G. 1977. The phytogeographic divisions of Amazonia and their influence on the selection of biological reserves. In *Extinction Is Forever*. New York Botanical Garden, New York. Pp. 195–213.

Putz, F. E., and M. A. Pinard. 1993. Reduced-impact logging as a carbon-offset method. *Conservation Biology* 7(4):755–757.

Reid, J. W., and R. E. Rice. 1997. Assessing natural forest management as a tool for tropical forest conservation. *Ambio* 25:382–386.

Rice, R. E., R. E. Gullison, and J. W. Reid. 1997. Can sustainable management save tropical forests? *Scientific American* 276(4):34–39.

Shirley, Danielle. 1997. Noel Kempff Mercado National Park, Bolivia. *Natural Assets* Fall:10–13.

Uhl, C., P. Barreto, A. Veríssimo, E. Vidal, P. Amaral, A. C. Barros, C. S. Jr., J. Johns, and J. Gerwing. 1997. Natural resource management in the Brazilian Amazon: An integrated research approach. *BioScience* 47:160–168.

USIJI 1996. "Joint Implementation Projects Selected: U.S. Companies Employ Innovative Solutions to Global Climate Change." Unpublished report of the U.S. Initiative on Joint Implementation, Washington, D.C.

Van Schaik, C. P., J. Terborgh, and B. Dugelby. 1997. The silent crisis: The state of rain forest nature preserves. In *Last Stand: Protected Areas and the Defense of Tropical Biodiversity*, edited by R. Kramer, C. P. van Schaik, and J. Johnson. Oxford University Press, Oxford.

Vincent, J. R. 1995. Timber trade, economics and tropical forest management. In *Ecology, Conservation and Management of Southeast Asian Rainforests*, edited by B. R. Primack and T. E. Lovejoy. Yale University Press, New Haven. Pp. 241–262.

Whitmore, T. C. 1988. *Tropical Rain Forests of the Far East*. Clarendon Press, Oxford.

World Resources Institute, United Nations Environment Programme, United Nations Development Programme, and World Bank. 1998. *World Resources 1998–1999*. Oxford University Press, New York.

PART IV

Mining and Conservation

11 Biodiversity Conservation, Minerals Extraction, and Development

Toward a Realistic Partnership

Alyson Warhurst
Kevin Franklin

The potential for minerals development to affect both the long-term health of ecosystems and biodiversity has long been a public concern. This stems from the 1972 Club of Rome report, *The Limits to Growth*, which predicted the imminent depletion of the Earth's nonrenewable resources, particularly fossil fuels and metals.[1] Despite such warnings, the discovery of new oil and mineral reserves, in conjunction with technical change and improved recycling, has alleviated fears of nonrenewable resource depletion (Warhurst 1998b). Since then, the environmental debate has shifted and is now focused on preventing the depletion and degradation of renewable resources such as water, air, land, and biodiversity. This is reflected in a growing emphasis on "sustainable development," which highlights the inter- and intragenerational inequities resulting from the degradation of renewable resources and the relationship between economic and social activities and environmental quality.[2] Central to the sustainable development agenda is the protection of ecosystems and biodiversity from any negative effects of industrial development.

It is partly as a result of the Convention on Biological Diversity, signed in Rio de Janiero, Brazil, in 1992, that governments are now implementing more stringent regulations regarding the sustainable use of natural resources. The development of national Biodiversity Action Plans (BAPs) and an increased requirement for prior environmental impact assessment (EIA), along with growing "voice of society" concerns focused on specific biodiversity hotspots, have obliged minerals companies to become more proactive in their approach to environmental management. While the Biodiversity Convention may showcase the need for the conservation of natural resources, it needs to be operationalized by both government and industry through the development of new management strategies.

Advances in theoretical and empirical research on biodiversity, rehabilitation, reforestation, environmental economics, social systems, ecotoxicology, management techniques, and so on have provided additional arguments for the reduction and prevention of further environmental damage. The multidisciplinary nature of much of this knowledge has made possible the development of innovative biodiversity conservation techniques and environmental management strategies in minerals development. A result of this "knowledge revolution" (Chichilnisky 1998) and innovative capacity (Warhurst 1994) has been that biodiversity conservation, minerals extraction, and sustainable development no longer appear to be incompatible.

This chapter analyzes the direct and indirect effects of minerals development on biodiversity, and the managerial and technological innovation strategies with which the minerals sector has responded to the expanding national and international regulatory climate and rising public concern. We explore these issues within the broader context of corporate social responsibility. We also emphasize the importance of EIA from the outset—in providing a comprehensive baseline data set and the criteria for ongoing monitoring—as a central aspect of the biodiversity conservation process. In addition, we emphasize that to be effective, such assessment requires innovative technologies and participative methods so that different bodies of knowledge, including, for example, indigenous traditional knowledge of values of biodiversity, are incorporated into the analysis from the outset and on an ongoing basis. The conclusions to this chapter highlight the need to focus on adaptive management strategies and innovative technologies, including biotechnology, in biodiversity conservation. Such issues are especially relevant in the light of the growing body of scientific knowledge on species, ecosystem dynamics and biodiversity conservation solutions.

Mining and Biodiversity: The Effects of Minerals Development

Minerals development may have significant effects on the integrity of both ecosystems and biological resources. Although minerals extraction per se requires a relatively small land area, the indirect effects of infrastructural development and the possible pollution associated with minerals development may be considerable. Such development can compromise the health and functional ability of ecosystems. These functions are responsible for a crucial part of human and environmental well-being through the provision of ecosystem goods and services. The latter include: air and water purification; weather ame-lioration; control of the hydrological cycle; generation and conserva-tion of fertile soils; dispersal and breakdown of wastes and nutrient cycling; control of many crop pests and disease vectors; crop pollina-tion; the provision of food; and a genetic library that may contribute to crop production, medicines, and industrial products (Faizi 1998; Vogt et al. 1997; Mooney et al. 1996). A study by Robert Costanza and others (1997) has estimated the value of seventeen major ecosystem services and functions as being in excess of US$33 trillion per year.[3] The extent to which the health and functional ability of ecosystems is disturbed by minerals development differs depending on the man-agement strategies or technologies being employed by the operator, as well as the nature of the local geology and geography and the ecosys-tem itself.

Land Use and Fragmentation

Among the most direct effects of minerals development on the health and function of ecosystems and biodiversity is its inherent use of land. The removal of biota, land, and minerals during the mining pro-cess may compromise ecosystem integrity through a loss of species or habitat or the destruction of watersheds. Among those impacts associ-ated with minerals development is the disruption, removal, erosion, or contamination of the soil environment. These may severely com-promise the ability of biodiversity to facilitate soil conservation (Faizi 1998). This loss may jeopardize continued agricultural productivity, mine site restoration programs, and the ability of ecosystems to pro-vide a host of additional goods and services.

For example, in addition to being the potential cause of erosion and crop loss, the disruption, contamination, or removal of the soil environment can have serious effects on the presence of soil mi-crobes. Of those organisms present in the soil, there are a series of fungal or bacterial symbionts that may form associations with forest plants and trees. Plants benefit from maintaining this symbiotic asso-ciation, which may provide: increased effective root surface area; in-creased availability of soil nutrients; increased availability of atmo-spheric nitrogen; increased heat and drought tolerance; increased tolerance to heavy metals; deterrence of infection by disease organ-isms; a food source for small mammals and humans; and ecosystem stability.

By compromising the integrity of the soil environment, minerals development threatens to disrupt the symbiotic associations that exist between such microbes and forest trees. The ability of forest trees to provide us with goods or services may be additionally compromised through their reduced ability to take up nutrients or resist environmental perturbations such as heavy metal contamination. The dynamic, interlinked nature of ecosystems suggests such a loss may also affect the human and animal components. While mine site restoration may go some way to replacing soil flora and fauna, it may take decades for plants, trees, and microorganisms to reestablish these symbiotic associations.

While desert and arctic biomes may be more vulnerable to ecosystem disruption because of their minimal functional redundancy, tropical biomes are often considered to be more in need of preservation, largely due to the high numbers of species (both endemic and nonendemic) that flourish in these environments. (Such systems, however, may display greater environmental resilience due to higher species numbers [C. Darwin 1859] and the compensatory nature of their interactions. This is reaffirmed by K. A. Vogt and others (1997), who note that an ecosystem is more resistant to perturbations if it contains a diversity of species, abiotic components, and functional types.)

Of the seventeen megadiversity countries noted in the following list (compiled by CI [1998]), which are home to tropical ecosystems and a great diversity of species, eight are on the list of emerging minerals markets compiled by *Mining Journal* (1996; starred). The importance of mineral reserves in these countries means they are also likely to be the focus of future minerals development, resulting in a range of possible environmental impacts.

Table 11.1 Megadiversity Countries and Emerging Minerals Markets

Megadiversity Countries	Emerging Minerals Markets
Australia	Argentina
Brazil	Chile
China	Peru*
Colombia	Brazil*
Democratic Republic of Congo	Indonesia*
Ecuador	Mexico*
India	Ghana
Indonesia	Bolivia
Madagascar	Philippines*
Malaysia	Venezuela*
Mexico	Zimbabwe
Peru	Namibia
Philippines	Kazakhstan
South Africa	Papua New Guinea*
Papua New Guinea	South Africa*
United States	Venezuela

While mineral extraction may have a direct impact on the landscape, the associated prior impacts of infrastructural development and the intrusion of human society into previously undisturbed environments has a potentially far greater effect on ecosystem integrity. Transportation corridors and utility infrastructures that accompany mine development fragment the landscape and facilitate human access to sensitive habitats and vulnerable species (Fleming 1998). Roads penetrating tropical forests may provide access to loggers, peasant farmers, ranchers, and plantation owners (Linden 1998). Such development may have serious impacts on both biodiversity and cultural diversity, especially when coupled with regional urban development resulting from increased economic activity associated with the mining project. While the direct effects of mining are localized, the indirect effects of regional development and pollution on both terrestrial and aquatic ecosystems may be relevant across a series of spatial scales, including the national and international level. For this reason an understanding of the interrelatedness of ecosystem functions and minerals development issues is instrumental in conserving biodiversity and achieving best practice.

Pollution Resulting from Mining Operations

In addition to the direct effects of minerals development on biodiversity, there are a series of further mining and mineral processing activities and waste dump integrity issues that may have indirect effects on ecosystems. These include acid rock drainage, tailings, and the seepage of cyanide, arsenic, lead, and oxides of carbon, sulphur, and nitrogen. Many of the effects on biodiversity could be prevented with the appropriate planning design and innovation; many such emissions can be naturally cycled within the ecosystem. Surplus concentrations, however, can cause drastic environmental effects. In addition to the excess in concentration, the media by which these emissions reach ecosystems plays a significant role in the spatial scale and how they might affect species and ecosystem function. This is nowhere more evident than in the Omai tailings incident (Guyana), which involved the discharge of over three billion litres of cyanide-contaminated waters into the Omai Creek and Essequibo River. The Guyana Legal Defence Fund (1998) notes reports of dead animals floating in the river, fish suffocating from the poison, and children suffering nausea. Similar tailings incidents such as those in Marcopper in the Philippines and Los Frailles in southern Spain have been fundamental in leading to a revision of national environmental and mining legislation, as well as investment in associated technologies and environmental management procedures.

Despite the wide range of information available on the effects of heavy metals, arsenic, cyanide, sulphuric acid, greenhouse gases, and so on, on ecosystem and human health, there is little information available on the cumulative effects of such impacts on the biological resource base. How such impacts may combine with the additional effects associated with landscape fragmentation and other anthropogenic disturbances is not yet fully understood. When taken in conjunction with seasonal and natural processes associated with ecosystems, such impacts may act to push the system beyond its natural state of fluctuation, perhaps leading to a loss of both ecosystem and species integrity. This loss will impact the direct use value of ecosystems as medicinal or agricultural goods sources as well as the less quantifiable functional, existence, and option values (Costanza et al. 1997).

Standards and agreements on the minimization of biodiversity loss and maintenance of ecosystem function were established at the U.N. Conference on Environment and Development (UNCED) in Rio de Janeiro in 1992. Signatories to these agreements have since produced national biodiversity action plans or strategies that may be instrumental in compelling environmental management, and specifically biodiversity conservation, within the minerals sector (South African Government 1998).

Managing Biological Resources in Minerals Development

An incomplete knowledge of biodiversity and the dynamic nature of living systems may explain some of the uncertainty in traditional environmental management systems. The establishment of goals for biodiversity conservation may be additionally constrained by incomplete taxonomic or functional data regarding the identity or role of species in ecological processes. An inability to predict ecosystem vulnerability or the economic implications of biodiversity loss further constrains the achievement of conservation goals. Attainment of such goals is further hindered by the dynamic, evolving nature of ecosystems. For these reasons adaptive management strategies have been noted as a potential mechanism for addressing the conservation of biological resources (Spellerberg 1991). The ability to adapt management strategies in response to the accumulation of new information is a vital aspect of biodiversity conservation. Adaptive management is especially useful when decisions need to be taken in the absence of data or where uncertainty is high.

To ensure that management strategies toward biodiversity conservation are implemented effectively, policymakers need an understanding of the social, economic, political, and cultural context in which they are implemented. Consequently, management plans may fail if they ignore the role of traditional knowledge and public concern about conservation. Such plans would also be in jeopardy if they failed to take issues of poverty and socioeconomic development into consideration. Effective biodiversity management is facilitated by a multidisciplinary, intersectoral approach. We examined such strategies hereafter in the context of a discussion of emerging best practices.

Emerging Best Practice

Companies often help conserve biological resources through rehabilitation and restoration practices that build on enhanced assessment and monitoring procedures. Although most of the new technologies used in minerals development do not evolve with the exclusive aim of conserving biodiversity, such use can be seen as a byproduct of recent innovation. We discuss examples of such technologies and practices hereafter. The accumulation and transfer of knowledge that results from both prior biophysical assessment and technology transfer sets in place the social and cultural infrastructure required for the sustainable use of biodiversity. Such advances may be especially relevant in terms of minerals development in remote regions with significant but vulnerable ecosystems.

Many dynamic firms foresee an evolution toward stricter environmental regulation and are making new investments in environmental management. They acknowledge that it will be to their competitive advantage to push environmental and technological frontiers forward. In the case of new projects, relatively free of the encumbrance of sunken investments in pollutant-producing obsolete technology, and with sufficient resources for research and development and technology acquisition, these firms have chosen either to develop cleaner process alternatives or to select new improved technologies from mining equipment suppliers (who are themselves busy innovating). Such firms are also likely to invest more in research aimed at enhancing biological and social resource integrity—that is, they are more "responsible." Increasingly, new investment projects are incorporating both improved economic and environmental efficiencies into production processes (Warhurst 1998c), not just in terms of new plants or equipment but also through the development of improved management and organizational practices.

Exploration

Much of today's mineral exploration and development takes place under the watchful eye of national and international governmental and nongovernmental organizations (NGOs). This oversight is especially relevant in remote regions harboring high levels of biodiversity and indigenous peoples. Many such regions, as noted earlier, have been identified as being both megadiversity countries and emerging markets for minerals development. One such venture has been the exploration of southeast Madagascar for titanium dioxide in the form of ilmenite and rutile mineral sands. The high number of endemic species found in Madagascar and in the mining path itself (Lewis Environmental Consultants 1992) has placed this development at the center of the international conservation agenda. Research suggests that minerals development in this region could have serious effects on the integrity of the ecosystem and the indigenous population. In examining the costs and benefits of mining, the operators acknowledge the requirement for further environmental and social assessments, together with technical studies. Such studies will be continuously reviewed within the context of the overall plans for the region and are to take place in line with a government-established process (Rio Tinto 1997). Minerals extraction will only take place following a comprehensive environmental, economic, and social appraisal and trade-off analysis. Such plans are best complemented by a complete set of environmental and social performance indicators, which will only be completed through investment in research, information, and empowerment of the local population. Although development of the region will have a significant impact on Madagascar's economy, such factors alone are no longer seen as being conclusive in the case of minerals development.

Innovation for Remediation and Reuse throughout the Mine Operation

Although many companies are moving toward pollution prevention through source reduction at the outset, the legacy of mining operations, in particular, has left many areas sufficiently contaminated to prohibit productive reuse (Warhurst 1998a). Regulations encouraging or requiring the rehabilitation of land for environmental, health, and commercial reasons have encouraged innovation in the field of cleanup technologies (Warhurst 1998b). Since any one company will have a portfolio of operational and closed properties (often on the same site), individual mining companies may use a combination of technologies for environmental management. In some circumstances it has proved possible to combine cleanup treatment operations with the recovery and reuse of saleable metals and minerals. Some liability provisions in the regulatory framework, however, can discourage remediation by increasing the potential risk for firms wishing to utilize innovative cleanup or waste remediation schemes (Tilton 1992).

After a period of using rather "static" technology, the mining and mineral processing industry is currently going through a phase of technical change, as dynamic firms develop new smelting and leaching technologies to escape economic as well as environmental constraints. This phase of technical change has been motivated by rapidly evolving environmental regulatory frameworks in the industrialized countries, reinforced by credit conditionality governing investment in the developing countries. Changing technological and environmental behavior in this context is evident particularly in the large North American and Australian mining firms and is increasingly becoming apparent in operations based in developing countries. (It seems that new operators and some dynamic private firms are changing their environmental practices most quickly while both state-owned enterprises and small-scale mining groups in developing countries continue, with some exceptions, to face technological and financial constraints regarding their capacity to change environmentally damaging practices.) Examples of some of these innovative technologies, which benefit biodiversity by reducing negative effects, follow.

Innovative Smelter Feedstocks

Regulatory limitations on waste disposal methods in the United States have generated some innovative approaches to the treatment of sludge wastes. The Resource Conservation and Recovery Act (RCRA) defines a set of hazardous wastes that can not be disposed of in landfills. During the cleanup of its Superfund site at California Gulch in Colorado, ASARCO was prevented by RCRA from disposing of metal-rich wastewater sludge on land. A search for alternative disposal methods led the company to develop a cost-effective procedure for utilizing sludge as a smelter feed stock. Not only did this innovative approach facilitate cleanup at the site and produce a solid waste that could be used as a lime flux replacement at ASARCO's lead smelting operations, but also it enabled the recovery of saleable metals and reduced ecosystem impact.

Liquid Effluents

Liquid effluents from mining operations have become a major concern of public groups, and this concern has translated into increased regulation of water quality impacts. Much public attention has focused on the possibility of cyanide escape from gold cyanidation operations. In response to increased scrutiny of the water quality impacts of these leaching operations, several innovative methods for destroying or recycling cyanide have been developed in recent years. INCO, for example, has commercialized a process that utilizes a copper catalyst in the presence of sulphur dioxide to produce oxidizing conditions sufficient to destroy cyanide. This technique has been applied at a number of operating sites across North America. Noranda has developed a similar process that is being used at Noranda's subsidiary, Hemlo Gold Mines' Golden Giant Mine. However, some potentially negative features remain in all chemical destruction processes. The reagents used are expensive and may have other as-yet-uncharted environmental impacts, including the residual precipitation of heavy metal cyanide complexes.

Cyanisorb

Cyprus Gold New Zealand developed a method of cyanide recovery that involves passing cyanide-containing wastes through a solution of sodium hydroxide or carbonate. The company had tested several existing cyanide destruction techniques for use at their new gold mine near Waitii in New Zealand's North Island. The technology allows for the removal and recycling of cyanide before it goes to the tailings pond. Not only does this help to solve environmental problems and facilitate future permitting, but also it economizes on the use of cyanide, a relatively expensive reagent, and improves the recovery of gold and silver. Recovery is normally greater than 90 percent, but the recovered hydrogen cyanide must be managed carefully due to its toxicity (Mitchell 1998).

Examples such as these tend to suggest that companies are not closing down, reinvesting elsewhere, or exporting pollution to less restrictive regulatory regimes in developing countries (Warhurst 1998b). Rather, they are adapting to environmental pressures by innovating, improving, and commercializing their environmental technology and management practices at home and abroad.

Rehabilitation and Decommissioning

While many governments, organizations, and mining companies strive to achieve best practice at new mine sites, there remains the challenge of the rehabilitation of older operations. In Chile the Ministry of Mines has commissioned a study on the potential revegetation and phytoremediation of soils polluted with heavy metals as a result of copper mining and smelting activities there (Danielson and Ginocchio 1998). This study takes local soil and waste conditions into account to help establish which plant species might be used in the revegetation process. Chile is also devising a formal mine closure planning system, in line with the legislation noted earlier. The national geology and minerals service has recently notified companies of steps that should be included in mine closure (Danielson and Ginocchio 1998).

While proactive international companies have submitted closure plans, this is seen partly as an effort to gain government approval and shield them against future government requirements or private claims for environmental damage. To an increasing extent, it is international insurance, guarantors, and lending bodies (e.g., Overseas Private Investment Corporation and Multilateral Investment Guarantee Agency) that dictate the need for environmental safeguards. The importance of financial drivers (Hughes and Warhurst 1998) in biodiversity conservation is exemplified through the withdrawal of private risk insurance at the Grasberg mine in Irian Jaya, Indonesia (Warhurst 1998b; Mealey 1996). Such developments have increased environmental awareness among both governments and minerals companies.

The decommissioning of Valdez Creek, Alaska, by Cambior Alaska stands as an excellent example of the increasing environmental awareness being demonstrated by mining companies and the role of federal and state legislation in driving such actions. The environmental impact of extraction was minimized through effective reclamation of the site, which entailed extensive landscaping, in-filling, and streambed recontouring. The 245-hectare property was awarded the U.S. Bureau of Land Management's Health of the Land Award and is now a wildlife haven and public recreation area (*Mining Journal* 1997). Such developments may both contribute to biodiversity conservation and preserve the functional ability of endangered ecosystems.

In addition to the ecosystem and biodiversity impacts of mining, some governments and national organizations have come to realize the occupational health effects of minerals development (Schaeffer et al. 1988; Costanza 1992; Haskell et al. 1992; Cairns et al. 1993; Rapport 1995). In doing so, these bodies recognize the tight link between social and ecological systems. Such issues are especially relevant in Australia, North America, and developing countries. Nicaragua, under the Convenio CentroAmericano–USA (CONCAUSA), has been involved in the support of research on the health issues of mercury contamination in small-scale mining activities (Eisen 1998). The International Development Research Centre of Canada (IDRC) is funding a research project, Environmental and Social Performance Indicators and Sustainability Markers in Minerals Development: Reporting Progress towards Improved Ecosystem Health and Human Wellbeing.[4]

National Legislation

National environmental legislation in the area of biodiversity conservation has been heavily influenced by the development of biodiversity action plans and strategies following the UNCED agreements. Although such legislation is often more advanced in industrialized nations, it is particularly relevant in developing countries, where, in the absence of a strong state and empowered stakeholders, it is argued that, especially where regulation is weakly developed or enforced, mining companies should develop their own models of environmental and social responsibility that go beyond acting within their more narrowly defined legal obligations to employees and shareholders (Warhurst 1998c). We briefly examine such issues here in a series of three examples.

Chile

Minerals development in Chile has, according to some observers, traditionally been oriented toward short-term economic growth, with negative implications for biodiversity conservation. Over the last fifteen years, Chile has come to appreciate the significance of the destruction of habitats and resources and the contamination of waters that occurred as a result of such activities. Minerals development in the region has been linked to the endangerment of tamarugo (*Prosopis tamarugo*) and cushion plant (*Azolrella laretia*) species (Danielson and Ginocchio 1998). The return of foreign companies in the 1980s and the recognition of the importance of the environment by those in the public sector led to the creation of the Environment Unit of the Ministry of Mines in 1991. Largely as a result of foreign investment and the concomitant evolution of environmental policy, a general law on the environment was developed. Established in 1994, this law aims to enforce mitigation, EIA, and adequate decontamination plans. The Environment Unit is also examining legislation on the mitigation of negative impacts at closed mines, abandoned tailings ponds, and smelter environs through revegetation (Danielson and Ginocchio 1998).

Namibia

Dynamic environmental legislation is also prevalent in countries harboring vulnerable yet valuable biological resources such as Namibia. Namibian mining licenses are currently subject to the approval of a long-term plan to safeguard the environment (Shekarchi 1996). This includes an EIA and management plan. The strenuous nature of such legislation is partly due to the semiarid and fragile Namibian environment. Reduced functional diversity within such systems makes them more susceptible to the impacts of minerals development. Scarcity of water has led mines to develop techniques for the minimization of water loss and use (Shekarchi 1996). In the light of such constraints the government is developing a national environmental management system. Such plans should help to ensure the preservation and conservation of ecosystem function and biodiversity.

The Philippines

Nowhere is the preservation of such functions more important than in megadiversity countries such as the Philippines. The Philippines has also been deemed one of twenty-five global biodiversity hotspots by CI (1998)[5] and was ranked ninth on *Mining Journal's* 1996 list of emerging markets (shown earlier). In this respect, it is home to significant biological and mineral resources. Mining earnings in the Philippines amount to 1.5 percent of the national gross domestic product and 4.1 percent of total exports (Disini 1997). The Mining Act (1995) created an attractive investment climate for foreign multinational corporations by permitting full ownership in large-scale exploration and development. It also led to the enhancement of environmental legislation through the subsequent revision of the Implementing Rules and Regulations (IRR) by the Department of Environment and Natural Resources (DENR). Such revisions emphasized environmental protection and participation of cultural minorities in ancestral lands affected by mining operations. The IRR stipulated that: mining companies must provide postoperation land management plans for open pits, waste dumps, and tailings dams; allocate 10 percent of the initial (capital) cost of a mine to environmental work; set aside 3–5 percent of mining and milling operating costs for an environmental protection program; and pay a new (higher) fine of US$2 per tonne of waste material spillage (Barker 1998). The development of such legislation was largely influenced by the Marcopper tailings incident of 1995 (Disini 1997). Only two minerals development applications have been approved since implementation; these include Western Mining Corporation's area in Cotabato, Tampakan, and Arimco's Didipio Project. Despite such legislation, fears remain, unsurprisingly, on the part of those in the Tampakan region that minerals development may lead to environmental damage. Such potential damage, they argue, would be in conflict with Article 2, section 16 of the 1987 Constitution, which guarantees the Filipino people the right to a "balanced and healthful ecology in accord with the rhythm and harmony of nature."

Transformation and Corporate Social Responsibility within the Minerals Sector

The changing national and international agenda on sustainable development and biodiversity conservation has been echoed by significant structural change within the international mining industry. Over the past decade, dynamic mining companies have begun to restructure their operations in response to new opportunities, arising from the liberalization of investment regimes for mining in many developing countries, economic liberalization, and the free flow of information, all of which have helped to open opportunities for companies to develop or acquire new production technologies and to respond to heightened environmental awareness and public scrutiny of their operations (Warhurst 1998b). This bundle of technological and organizational changes has the potential, if effectively managed, to contribute to economic growth and improved environmental and social performance in developing countries. Since 1989 over seventy-five countries have liberalized their investment regimes for mining. Economic and political reforms have opened up new opportunities to the international mining industry in areas that were formerly closed, since economic and political risks were sufficiently high to deter prudent investment. Private investment flows to developing countries have increased in response to these opportunities, with mining playing a significant role as a proportion of the total direct foreign investment. For example, in 1995, direct foreign investment in developing countries was valued at $90 billion while capital expenditures on mining alone were estimated at $20 billion for the period 1995–2000 (Warhurst and Lunt 1997).

This transformation of investment regimes and patterns of investment flows is occurring at a time of significant technological change within the mining industry, as firms respond to increasing market and regulatory pressures. Incremental improvements in process control and optimization, or the application of existing technologies at increasing scales of operation to capture greater efficiencies, have proved fundamental to maintaining the competitiveness of major mineral producers. Not only have innovations in processing technology improved productivity and efficiency but, by improving process control, increasing recovery rates, and reducing waste, several key processing innovations have enabled firms to combine gains in competitiveness with improved environmental performance. Such performance enhancement, when coupled with strategic environmental management and financial drivers, could have positive effects on the conservation of biodiversity in developing countries. Direct foreign investment through joint ventures with state-owned firms and/or newly privatized entities in developing countries may, under certain conditions, provide an effective vehicle for the transfer of conservation technologies, generating improvements in production efficiency and environmental performance and enhancing long-term shareholder value.

However, while mining companies during this recent phase of globalization have contributed to improved social development, by providing jobs, paying taxes, building an industrial base, enhancing efficiency, earning foreign exchange, and transferring technology, they have also been linked publicly to interference in sovereign affairs, deepening disparities in wealth, poor labor conditions, corruption, transfer-pricing, pollution incidents, damage to biodiversity, health and safety problems, and the disrespect of human rights. A growing and more effectively enforced body of regulation in this field should also have a positive impact on environmental integrity. When cou-

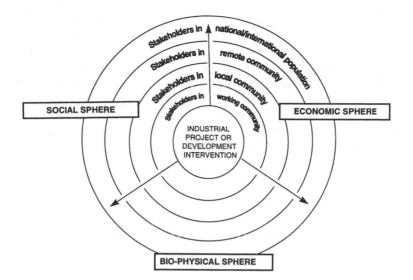

Figure 11.1 Stakeholder spheres.

pled with technological innovation, EIA, and effective monitoring programs, biodiversity conservation becomes more of a reality.

Evidence suggests that mining companies will increasingly be obliged through public pressure to make pledges to different stakeholders and set targets relevant to environmental and social responsibility within the biophysical, social, and economic spheres. Figure 11.1 shows a model in which the mining project can be considered the input and the health and well-being of affected stakeholders are outputs. This model links social, environmental, and economic performance by suggesting that corporate social responsibility should not be considered independently of effects over time within the biophysical and economic spheres. This new approach to corporate environmental and social responsibility requires a conceptual shift in the understanding of the positive and negative effects of multinational corporations on environmental and social development, from a position oriented to legal rights to one oriented to social responsibility. While it can be argued that multinational corporations are only legally responsible to employees and shareholders, the question remains whether such corporations have some responsibility for advancing social and environmental goals, given the transboundary nature of their operations and the lack of provision of public services by governments in many remote areas. Particularly in developing countries, where disparities in power and information clearly exist, the crucial issue is whether their profit-making nature and the fact that their rights are enshrined in current national and international law frees multinational corporations from moral obligation and social responsibility. A growing part of society views such a narrow interpretation of rights as cynical and the abrogation of broader responsibilities as opportunistic. In order to address such concerns, promote best environmental practice, conserve biodiversity, and implement the recommendations of both the Convention on Biological Diversity and other international environmental agreements, companies should recognize certain underlying principles common to these agreements, as follows (Grubb et al. 1993).

1. The need for a "bottom-up" approach towards management—placing additional emphasis on the role of major groups such as indigenous peoples and NGOs.
2. The importance of adequate information—possibly through a well-planned management strategy and the effective use of available tools e.g. EIA, Biodiversity Impact Assessment etc.
3. The need for adequate cross-cutting institutions—collaboration between business, industry, NGOs, governmental organizations, universities etc. to facilitate the most effective management strategy for sustainable use of natural resources e.g. biodiversity conservation.
4. Recognition of the complementarity between regulatory approaches and market mechanisms for addressing development and environmental needs.

Future Challenges

Dynamic companies have responded to societal perceptions and other nonregulatory and regulatory drivers by improving environmental performance as part of their search for a competitive edge. In many cases environmental performance and standards continue to improve. Companies are now beginning to operate within the regulatory standards and guidelines of their home country or of international organizations such as the World Bank and bilateral development agencies in countries where less stringent legislation or weaker enforcement would otherwise apply. This suggests that a commitment to anticipating and minimizing environmental impacts from the outset is growing and that competitiveness and enhanced environmental performance are not mutually exclusive. Similarly, it is becoming increasingly clear that technological or managerial incompetence is not compatible with sustainable development. Limitations arising from initial choices of technology can potentially undermine even the best efforts in environmental and human resource management and stakeholder consultation.

A firm's responses to environmental pressures and its capacity to contribute positively to conserving biodiversity also depends on the nature of its operations in terms of: (1) the mineral involved; (2) the level of integration of mining and processing activities; (3) the stage in the investment and production cycle at which its mineral projects operate; and (4) the internal economic and technological dynamism of the firm (i.e., whether the firm has the financial, technical, and managerial capabilities to be an innovator) (Warhurst 1998c).

Dynamic firms, which have new project development plans in place, are in the most favorable position to invest in more environmentally sound alternatives or to raise the capital to acquire new technologies that have the potential to be more economically efficient as well as environmentally less hazardous. Some mining firms have even pushed technology beyond the bounds of existing regulations. As a consequence, these firms are supporting tighter and more globally consistent environmental regulations, because they are able to utilize their new environmentally sound technologies and management systems to meet stricter standards and obtain a competitive advantage over less technologically advanced firms. Such firms are also in a stronger position to respond to supply-chain pressures, such as the requirement of metal product retailers and the construction industry for more environmentally friendly production methods.

It is no longer acceptable for corporations to cause any environmental and social disruption in favor of short-term economic gain. The technology and management expertise now exist to prevent or manage any level of pollution inside or outside the mining project boundaries. Comprehensive analysis of minerals development now needs to factor in, from the outset, the complex, multidimensional nature of biophysical, social, and economic systems. Minerals development should be compatible with long-term ecological and social stability The challenge to biodiversity conservation in new mining projects lies in the development of an effective yet comprehensive means of understanding, monitoring, and managing environmental change from the outset.

Environmental regulation is here to stay and is sure to become more widely adopted, more stringent, and better enforced. Financial drivers will be stricter, public pressure more exacting, and the penalties more substantial. As a result, it will not be those firms that avoid environmental control that win a greater share of the metals market. It will be those firms that were ahead of the game, that played a role in changing the industry's production parameters and used their innovative capabilities to improve their competitive advantage and ensure the sustainable use of both renewable and nonrenewable resources. In ensuring the sustainable use of biological resources and cultural diversity, such firms will be contributing positively to safeguarding their security, and that of society, in perpetuity.

1. Meadows et al. (1972). The model projected that if the world economy continued to expand over the following fifty years as the economy of the United States had over the fifty previous years, then the entire system would collapse through the exhaustion of the supply of crucial materials and pollution from the massive increase in materials-intensive industrial production.
2. The 1987 Brundtland Report of the World Commision on Environment and Development defined sustainable development as "development that meets the needs of the present without compromising the ability of future generations to meet their own needs." Implicit in this definition are the following ideas: (1) environmental matters should be incorporated into economic policies; and (2) economic development should promote intergenerational and geographical equity. Sustainable development has been discussed in detail by many academics including: Jacobs (1991), Pearce et al. (1989), Pezzey (1989), and Redclift (1987).
3. Those ecosystem services and functions used in this study include: gas regulation, climate regulation, disturbance regulation, water regulation, water supply, erosion control and sediment retention, soil formation, nutrient cycling, waste treatment, pollination, biological control, refugia, food production, raw materials, genetic resources, recreation, and cultural services. Ecosystem "goods" were included with ecosystem services.
4. The objective of this project is to promote environmentally sensitive practices in the development of nonrenewable natural resources and to contribute to the development of mechanisms and methodological tools that enhance the well-being of communities affected by mining projects. This project is being undertaken by the Mining and Energy Research Network at the University of Warwick, United Kingdom, Tata Energy Research Institute (TERI), in New Dehli, India, and Institut National del environnement Industriel et des Risques (INERIS), in Bogota, Colombia.
5. CI first conducted the "hotspot analysis" in 1990. It was reviewed and updated in 1996. Designation as a biodiversity hotspot is based on two criteria: the number of endemic species an ecosystem contains and the degree of threat it faces. Despite occupying less than 2 percent of the Earth's land area, the hotspots contain more than 40 percent of terrestrial species that exist nowhere else (endemics).

REFERENCES

Agency for Toxic Substances and Disease Registry (ATSDR). 1989. "Toxicological Profile for Arsenic." Draft. U.S. Public Health Service, U.S. Department of Health and Human Services, Atlanta, Ga. http://atsdr1.atsdr.cdc.gov
———. 1992."Toxicological Profile for Mercury." Draft. U.S. Public Health Service, U.S. Department of Health and Human Services, Atlanta, Ga. http://atsdr1.atsdr.cdc.gov
———. 1993. *Toxicological Profile for Fluorides, Hydrogen Fluoride and Fluorine*. U.S. Public Health Service, U.S. Department of

Health and Human Services, Atlanta, Ga. http://atsdr1.atsdr.cdc. gov

Barker, J. 1998. *Social and Economic Policies for Minerals Development in Australia and the Philippines*. Mining and Environment Research Network, Working Paper no. 112. School of Management, University of Bath, England.

Cairns, J., Jr., P. V. McCormick, and B. R. Niederlehner. 1993. A proposed framework for developing indicators of ecosystem health. *Hydrobiologia* 263:1–44.

Chichilnisky, G. 1998. The knowledge revolution. *Journal of International Trade and Economic Development* 7(1):39–53.

Conservation International. 1998. *Global Biodiversity Hotspots*. Conservation International, Washington, D.C. http://www.conservation. org/

Costanza, R. 1992. Towards an operational definition of ecosystem health. In *Ecosystem Health: New Goals for Environmental Management*, edited by R. Costanza, B. G. Norton, and B. D. Haskell. Island Press, Washington, D.C.

Costanza, R., R. d'Arge, R. de Groot, S. Farber, M. Grasso, B. Hannon, K. Limburg, S. Naeem, R. V. O'Niell, J. Paruelo, R. G. Raskin, P. Sutton, and M. van den Belt. 1997. The value of the world's ecosystem services and natural capital. *Nature* 387(6230): http://www.nature.com/

Danielson, L. J., and R. Ginocchio. 1998. Chile: restoration challenge . . . and mine closure. *Mining Environmental Management*, March 1998, pp 7–12.

Darwin, C. 1859. *The Origin of Species*.

Disini, A. F. 1997. *Philippines 1997 Update*. Mining Journal Interactive Country Guides, London, England.

Eisen, L. B. 1998. *Social and Economic Policies for Minerals Development in Nicaragua*. Mining and Environment Research Network, Working Paper Series, no. 110. School of Management, University of Bath, England.

Faizi, S. 1998. An assessment of the economic benefits of biodiversity in Saudi Arabia. *Natural Resources Forum* 22(1):63–66.

Fleming, S. 1998. *Mining and Conflicts Relative to Biodiversity in Latin America: Recommendations for Industry*. Mining and Environment Institute, Queens University, Ontario, Canada.

Franklin, K. C. 1997. Biodiversity indicators and minerals development. M.Sc. dissertation, University of Bath, England.

Grubb, M., M. Koch, A. Munson, F. Sullivan, and K. Thomson. 1993. *The Earth Summit Agreements: A Guide and Assessment*. Earthscan Publications, London, England.

Guyana Legal Defense Fund. Accessed 1998. *The Omai Mine Spill*. http://www.geocities.com/RainForest/2651/

Haskell, B. D., B. G. Norton, and R. Costanza. 1992. What is ecosystem health and why should we worry about it? In *Ecosystem Health: New Goals for Environmental Management*, edited by R. Costanza, B. G. Norton, and B. D. Haskell. Island Press, Washington, D.C.

Hughes, N., and A. Warhurst. 1998. *The Financial Drivers of Environmental and Social Performance*. Mining and Environment Research Network, Working Paper no. 142. School of Management, University of Bath, England.

Jacobs, M. 1991. *The Green Economy.* Pluto Press, Concord, Mass.

Lewis Environmental Consultants. 1992. *Madagascar Minerals Project EIA Study.* Pt. 1. Australia. Friends of the Earth, London.

Linden, E. 1998. Smoke signals: Vast forest fires have scarred the globe, but the worst may be yet to come. *Time,* June 22, 1998.

Marcus, J. A., ed. 1997. *Mining Environmental Handbook: Effects of Mining on the Environment and American Environmental Controls on Mining.* Imperial College Press, Singapore.

Meadows, D. H., D. L. Meadows, J. Randers, and W. W. Behrens. 1972. *The Limits to Growth: A Report for the Club of Rome's Project on the Predicament of Mankind.* Universe, New York.

Mealey, G. A. 1996. *Grasberg.* Freeport McMoran Copper and Gold, New Orleans, La.

Mining Journal. 1996. Emerging markets supplement. *Mining Journal* 327(8398).

————. 1997. *Alaska.* Country supplement. *Mining Journal.* 329(8456).

Mitchell, P. B. 1998. The role of waste management and prevention in planning for closure. In *Planning for Closure: Best Practice in Managing Ecological Impacts from Mining,* edited by A. C. Warhurst and L. Noronha. Lewis, New York.

Mooney, H. A., J. Lubchenco, R. Dirzo, and O. E. Sala. 1996. Biodiversity and ecosystem functioning: Basic principles. In *Global Biodiversity Assessment,* by UNEP. Cambridge University Press, Cambridge, England.

Pearce, D., A. Markandya, and E. B. Barbier. 1989. *Blueprint for a Green Economy.* Earthscan, London.

Pearce, F. 1987. Acid rain. *New Scientist: Inside Science* 116(1585):.

Pezzey, J. 1989. *Economic Analysis of Sustainable Growth and Sustainable Development.* Working Paper no. 15. Environment Department, World Bank, Washington, D.C.

Rapport, D. 1995. Ecosystem health: More than a metaphor? *Ecosystem Health* (Special Issue of Environmental Values) 4(4):287–311.

Redclift, M. 1987. *Sustainable Development: Exploring the Contradictions.* Methuen, London.

Rio Tinto. 1997. *Health, Safety and Environment Report 1996.* Rio Tinto.

Schaeffer, D. J., E. E. Herricks, and H. W. Kerster. 1988. Ecosystem health. Pt. 1. Measuring ecosystem health. *Environmental Management* 12(4):445–455.

Shekarchi, E. 1996. *The Minerals Industry of Namibia.* Minerals Information, U.S. Geological Survey, Washington, D.C. http://www.usgs.gov/

South African Government. 1998. *Green Paper on Mineral Policy of South Africa.* Available on the internet at http://www.polity.org.za/govdocs.

Spellerberg, I. F. 1991. *Monitoring Ecological Change.* Cambridge University Press, Cambridge, England.

Tilton, J. E. 1992. *Mining Waste, The Polluter Pays Principle, and U.S. Environmental Policy,* Working Paper 92-8, Department of Mineral Economics, Colorado School of Mines, Golden, CO.

U.S. Environmental Protection Agency. 1989. *Health Issue Assessment: Summary Review of Health Effects Associated with Hydro-*

gen Fluoride and Related Compounds. EPA/600/8-89/002F. Environmental Criteria and Assessment Office, Office of Health and Environmental Assessment, Office of Research and Development, EPA, Cincinnati, Ohio.

Vogt, K. A., J. C. Gordon, J. P. Wargo, D. J. Vogt, and collaborators. 1997. *Ecosystems: Balancing Science with Management.* Springer-Verlag, Berlin.

Warhurst, A. 1994. The limitations of environmental regulation in mining. In *Mining and the Environment: International Perspectives on Public Policy,* edited by R. G. Eggert. Resources for the Future, Washington, D.C.

Warhurst, A. 1998a. *The Environmental Management Strategies of Mining Companies: An Overview of the Issues.* Case Studies of Corporate Responses to Environmental and Social Challenges, Mining and Environment Research Network, Working Paper no. 125. School of Management, University of Bath, England.

———. 1998b. Environmental regulation, innovation and sustainable development. In *Environmental Management and Sustainable Development: Case Studies from the Mining Industry,* edited by A. Warhurst and L. Noronha. International Development Research Centre Ottawa, ON, Canada.

———. 1998c. Mining, mineral processing and extractive metallurgy: An overview of the technologies and their impact on the physical environment. In *Planning for Closure: Best Practices in Managing the Ecological Impact of Mining,* edited by A. Warhurst and L. Noronha. Lewis, New York.

Warhurst, A., and Lunt, A. 1997. *Corporate Social Responsibility: A Survey of Policy, Research and Consultancy Activity.* World Business Council for Sustainable Development (WBCSD) Report. University of Bath, England.

12 Mining Industry Responses to Environmental and Social Issues

Gary Nash

Globalization and sustainable development have become agents of change for society at large, including the nonferrous metals industry. They bring both new investment opportunities and raised public expectations about the role of corporations in the economic and social fabric of society. The first part of this chapter situates the nonferrous metals industry in the context of recent international developments and describes changes in corporate culture that are taking place in response to new world circumstances. The second part introduces the International Council on Metals and the Environment (ICME) and describes its work and that of its member companies in promoting environmentally and socially sustainable economic development in the field of nonferrous metals. (The various principles promoted and activities undertaken are of equal relevance to both industrialized and developing countries.)

Recent International Trends

Globalization is changing the world's way of doing business. Centrally planned economies are being replaced by market-based economies and democratic structures. Many countries that previously spurned foreign investment now seek it to stimulate their economies. Throughout the world, barriers to the movement of capital, trade, people, goods, and services are coming down. Globalization provides industry with a wealth of business opportunities and a more secure investment climate.

Meanwhile, the 1992 Earth Summit and the earlier publication of *Our Common Future* (commonly referred to as the Brundtland Report) by the World Commission on Environment and Development in 1987 have popularized the concept of sustainable development and brought environmental issues forward on the international agenda. The Brundtland Report defined sustainable development as "development that meets the needs of the present without compromising the ability of future generations to meet their own needs." The report also dealt with the disparity of wealth in the world and recognized the importance of economic development in alleviating environmental stress and improving living standards in developing countries.

These concepts of sustainable development now provide the foundation for the policies of many nations. They are also beginning to raise public expectations about the role of corporations in modern society. Corporations are no longer seen as entities operating solely for the benefit of shareholders within the political and regulatory framework established by the state. Rather, they are increasingly expected to assume greater responsibility for the economic and social impacts of their activities and to work with outside groups to improve the quality of life within the communities in which they operate.

Yet this new role for the private sector is not without skeptics. Nongovernmental organizations (NGOs) have questioned transnational corporations' commitment to environmental performance and good labor practices and their respect for the rights and interests of stakeholders other than their own shareholders. Worldwide television news coverage and the Internet allow real or perceived lapses in corporate behavior—whether poor environmental performance, worker exploitation, adverse impacts on traditional cultures and peoples, or weak business ethics—to be communicated rapidly to all parts of the globe. As a result, transnational companies are finding themselves more and more under the public microscope. Openness and transparency are therefore vital if industry is to reassure the skeptics and build the confidence and trust of the public and other stakeholders.

The Nonferrous Metals Industry

Few sectors have felt the impact of these recent international developments as much as the nonferrous metals industry. Globalization has opened doors of opportunity on a scale that was unimaginable just a few years ago. Many mineral-rich countries now see the development of a prosperous mining industry as an opportunity to stimulate overall economic activity and raise living standards. At the same time, NGOs are raising flags of caution, citing the intrusive nature of mining and warning of damage to the environment and indigenous cultures if proper management practices are not followed. Contrary to the views of some of its critics, however, the nonferrous metals industry can contribute to environmentally and socially sustainable economic development. Moreover, corporate cultures are becoming more sensitive to broader environmental and social responsibilities and are creating new progressive approaches within individual companies and at the industry level.

The Economic Dimension

Everywhere in the world, the progress and prosperity of individuals, communities, and societies depend to a certain degree on the economic production and availability of a broad range of metals. The physical and chemical properties of metals make them ideally suited for a wide range of industrial and social applications, including new uses in the electronics, computers, aerospace, medicine, and environmental management industries. This fundamental fact of life is not expected to change in the years to come. Indeed, population growth and improvements in the quality of life, especially in developing countries, will require new supplies of metals.

In response to the world's growing demand, the mining industry is seeking high-quality ore bodies in all parts of the globe. The discovery of deposits and their subsequent development can be powerful engines for advancing economic growth and social progress both in the developing world and in the more remote areas of industrialized countries, where there are few other industrial development options. International mining companies can help contribute to the creation of the social, human, and investment capital necessary to the future development of the societies in which they operate, *subject to the appropriate policies of host governments.*

Another important factor that is often overlooked is that metals can be recycled and reused almost indefinitely with major savings in energy and waste management costs. This makes metal-containing products a superior environmental choice among competing materials. Indeed, metals are the only resource that, once extracted, can lend themselves to infinitely continued use, thereby minimizing the future drain on the world's natural resources while providing for the needs of present and future generations.

The Social Dimension

Corporations do not operate in a social vacuum. The need to reach out to local mining communities and to society at large, in both the industrialized and developing worlds, has given rise to new sensitivities toward the social dimension of industry activities.

Mining companies are usually the major source of income and employment in the regions in which they operate. Moreover, mining operations in remote areas usually start with the development of basic physical and economic infrastructures that come to depend on the growth of these operations and on the indirect stimulation of other economic activities. These operations also contribute to the development of a region's social infrastructure (e.g., schools, hospitals, etc.). This position confers special responsibilities on the company to develop close relations with local communities in promoting the economic and social vitality of a whole area and in addressing any concerns that may arise about the impact of the mining operations.

Companies operating in close contact with indigenous peoples need to address a special set of issues (for example, disputes over land claims or preservation of traditional cultures and lifestyles). More and more, these companies are attempting to ensure that indigenous peoples derive greater benefit from resource development through direct participation in the workforce, the sourcing of goods and services, and the provision of training facilities. Social issues especially can dominate in developing countries. International companies must often walk the fine line between showing respect for local customs, traditions, and institutions and being an important agent for technology transfer and social progress. Companies are in an excellent position to provide the technical expertise and assistance needed to build up scientific infrastructure as well as the capabilities of local regulatory regimes. At the same time, however, they must respect the sovereign rights of national governments to manage their own natural resources on behalf of the people.

The Environmental Dimension

While the contributions of metals can be readily recognized by society, in more recent years public demand has been growing that the pursuit of economic development be compatible with environmental and social values. These demands have been transferred to the political agenda in the form of new legislation and regulatory requirements. Industry has responded accordingly. In the area of occupational health, for example, process technology has been introduced, and exposure to potential work hazards has been minimized through better worker training and awareness programs. Medical surveillance programs have been established to monitor the health effects of worker exposure to certain hazardous substances.

Regarding the environment, considerable progress has been achieved as a result of major investments by industry in cleaner processing technologies, rehabilitation efforts, effluent controls, and pollution prevention. Most large mining and metallurgical companies have adopted corporate environmental policies and implemented environmental management systems to ensure that environmental protection is a corporate priority and that environmental concerns are integrated into all aspects of corporate activities, from exploration to environmental impact assessment and from mine development to closure and rehabilitation of mine sites. Environmental management systems ensure compliance with corporate policies and regulatory requirements through regular monitoring, evaluation, and auditing of a corporation's environmental performance, including the establishment of emergency procedures. They also provide for corrective measures and continual improvement in a company's environmental performance. Careful planning and personnel training are integral elements of successful environmental management systems.

Processing and control technology and risk management techniques have evolved considerably in recent years. Through corporate commitment and the strict application of sound planning and management techniques, disturbances to the environment can be kept to a minimum.

The New Corporate Culture

The nonferrous metals industry continues to operate on an international pricing system, which means that prices are the same for all producers regardless of cost. It is therefore a cost-driven industry, one that strives to produce at the lowest possible cost in order to remain competitive and to attract investment capital. Notwithstanding these pressures, recent international developments have combined to create a dynamic, competitive industry that is also increasingly aware of its broader economic and social responsibilities. Sound environmental management, ethical business practices, and proper attention to community development are now seen as essential elements of good management and have been given high corporate priority in most major mining and metallurgical companies.

Many companies are also becoming more open and transparent. By sharing information with governments and local communities on the environmental aspects of their activities, and by participating actively in public dialogue on environmental issues and the development of policy prescriptions, the industry is establishing trust and building confidence in its activities. As a result, the industry is frequently asked to provide its technical expertise in the development of more results-oriented and scientifically based public policies. In some cases, the industry has been challenged to set its own targets for environmental improvement and to take the necessary action needed to achieve those objectives.

Finally, the competitive nature of the industry itself is shaping corporate cultures in ways that were totally unexpected even a few short years ago. More and more, industry is sharing information on new environmental technologies and "best practices." Because the industry is only as strong as its weakest member, companies have a vested interest in exchanging experiences on environmental issues and developing industry-wide codes of conduct. To facilitate this process, the world's leading producers of nonferrous and precious metals came together in 1991 to establish ICME.

The ICME's Purpose

Established to promote the development and implementation of sound environmental and health policies and practices in the production, use, recycling and disposal of metals, ICME is an international NGO with member companies from six continents that together represent a major portion of the world's production of nonferrous and precious metals. Members benefit from pooling expertise, exchanging information, and contributing to international discussions on important environmental and health issues. ICME also provides, on behalf of its member companies, a focal point for addressing environmental and health issues and policies of a generic nature at the international level. As part of this function, ICME has either official status with, or participates in the work of, specialized agencies of the United Nations, such as the U.N. Environment Program (UNEP), and other intergovernmental organizations, such as the Organization for Economic Cooperation and Development (OECD). In addition, ICME cooperates with a number of international business and nongovernmental organizations.

ICME's Environmental Charter

As a first major step in pursuing a clear and common direction on environmental matters, members adopted the ICME Environmental Charter, which includes an introductory philosophical statement and establishes a number of guiding principles—in the areas of product stewardship, environmental stewardship, local community responsibilities, and communication—which guide ICME's communications program (for text, see the end of this chapter). ICME believes that its Charter inspires other companies in the industry to give a high priority to good environmental practices.

ICME Activities

Information exchange, training, capacity building, and international cooperation are fundamental to furthering the Charter's objectives; accordingly, ICME serves as an information center. ICME's quarterly newsletter regularly publishes examples of environmental achievements on the part of individual member companies. It also organizes both internal workshops for its members and international seminars and conferences for industry at large and government authorities, NGOs, and others. ICME's conferences are often organized in cooperation with other international organizations. ICME's internal workshops have kept members abreast of environmental and social issues with implications for corporate practices and policies. Workshop topics have included environmental management systems, risk assessment and management, and mining and indigenous peoples. ICME's conferences and seminars have resulted in an increased awareness of environmentally sound mining practices among a wider audience, including government authorities and NGOs. Notable events organized to date include:

- The International Conference on Development, Environment and Mining in Washington, D.C. (June 1–3, 1994), cosponsored by the World Bank, UNEP, the United Nations Commission for Trade and Development (UNCTAD) and ICME and chaired by Henrique Cavalcanti, minister of the environment of Brazil;
- A seminar on environmental management in Almaty, Kazakhstan, cosponsored by the Kazakh Ministry of Ecology and Bio-Resources and ICME at the request of the Government of Kazakhstan, which was seeking solutions for the environmental problems associated with its mining and metals sector; and
- The International Workshop on Managing the Risks of Tailings Disposal, held in Stockholm, Sweden, and cosponsored by ICME and UNEP with the support of the Swedish International Development Agency. A number of recommendations were made, many of which are being pursued in 1998.

ICME Publications

ICME also operates an extensive publications program, which provides a means for disseminating information on subjects related to the health and environmental effects of the mining and production of nonferrous metals. One such publication, *ICME-UNEP Case Studies Illustrating Environmental Practices in Mining and Metallurgical Processes,* contains examples of Environmental Management Systems, project approval procedures, sound operations, techniques, training, and education. ICME has also published a state-of-the-art report on the recycling of nonferrous metals that provides comprehensive coverage of all aspects of recycling, including the properties of nonferrous metals, their uses, and the processing and economics of recovery. Other ICME publications address topics such as: endocrine disruptors, the use of cost-benefit analysis in chemicals management, risk assessment techniques, and hazard identification in the aquatic and terrestrial environments, among others.

ICME's Impact

ICME's activities have had a clear impact on the policies and practices of its member companies. A number of companies have adopted environmental policies based on the principles enunciated in ICME's charter. ICME's environmental and product stewardship principles are often cited in member companies' annual reports and environmental reports.

More and more members are responding to ICME's technical work. Some companies have adopted ICME's recommended approach to gathering data on occupational exposures in order to overcome existing database deficiencies. ICME technical workshops have also sensitized companies to various issues—for example, tailings management—and have resulted in improvements in corporate environmental management systems and practices. Social issues addressed in ICME workshops and the recently adopted Community Principles are beginning to be reflected in corporate policies and practices.

Industry Voluntary Initiatives

In recognition of the limitations of command-and-control mechanisms, governments are looking at other approaches to environmental improvement to supplement traditional regulatory regimes. Voluntary action by industry is one approach that allows companies to innovate and determine the best way for improving environmental performance based on their own unique circumstances and local environmental conditions. Examples of the various types of voluntary initiatives taken by ICME member companies and industry at large are outlined hereafter. While these examples are drawn from the United States, Canada, and Australia, many of them are applicable to other countries of the world, particularly those countries in which major transnational companies have operations.

Industry Codes of Practice

The mining industry in several countries has moved to develop codes of practice, whose purpose is to provide a framework to guide mineral companies, wherever they operate, toward effective environmental strategies for mineral development, from initial exploration to closure and final rehabilitation. They also provide for progressive improvement in environmental performance and increased industry accountability through public environmental reporting and public participation in decisions affecting where and how mineral projects proceed.

In Australia this approach has achieved an excellent response. In recognition of community concerns about the environmental performance of the minerals industry and in response to calls from the conservation lobby for consistent standards of environmental performance, in 1995 the Minerals Council of Australia developed its Minerals Industry Code for Environmental Management. To date, companies accounting for more than 85 percent of Australia's mineral production have signed on. Further work is required to increase the participation of smaller companies, and the Minerals Council and state minerals associations are actively working to achieve this.

The Australian code (see the text at the end of this chapter) is voluntary and open to all mining and minerals companies. However, the Code applies to all of the activities of a signatory company whether in Australia or overseas. While the Code does not prescribe specific environmental practices at mining and mineral processing sites, it does set out key principles for environmental management that allow signatories to progressively improve performance. Specific guidance notes are also being prepared to assist with implementation.

The contribution of the Australian mining sector to biodiversity conservation was also recently recognized in the Australian government's first national report to the Convention on Biodiversity. The baseline research undertaken by the mining industry as part of project approvals and management processes comprises an important proportion of the biological research work being done in Australia and represents a significant contribution to basic taxonomic, biological, and ecological information and understanding. The mining sector is also involved in species protection and enhancement programs both on and off site in an effort to manage and ameliorate the impacts of various mining projects on Australia's biodiversity. The industry also supports research programs that contribute to the understanding of the ecology, biology, or taxonomy of native species as well as the management, recovery, and rehabilitation of ecosystems and species populations.

Voluntary Industry Emission-Reduction Programs

Other types of initiatives include voluntary industry emission-reduction programs. Two notable examples include Canada's Accelerated Reduction/Elimination of Toxic Substances (ARET) program and the "33/50" initiative of the Environmental Protection Agency (EPA) of

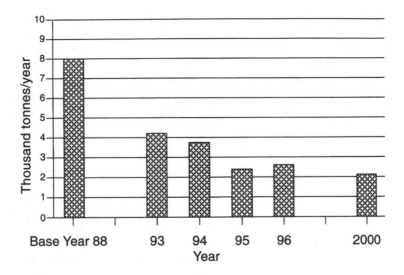

Figure 12.1 Tonnes of major ARET-listed substances released during the year by the mining industry ARET participants.

the United States. Established in the early 1990s, these multistake-holder programs seek voluntary reductions, on the order of 50 to 90 percent, in environmental releases of identified substances. One indication of the success of these programs is that participating nonferrous metals companies are ahead of schedule in meeting the targets (see Figure 12.1).

The Mine Environmental Neutral Drainage Program

Another notable voluntary initiative is the Mine Environmental Neutral Drainage (MEND) program. Sponsored by the Canadian mining industry, the government of Canada, and several provincial governments, MEND was established in the early 1980s in order to develop and apply new technologies to prevent and control acid drainage. Before its establishment, both industry and governments knew of the existence of acidic drainage but had no clear idea of how to solve the problem. MEND has since provided an increased and more sophisticated knowledge base that has made it possible to manage the complexity of acidic drainage. Of particular importance was the development of a common understanding among participants, which has allowed partners to take actions with greater confidence and to gain multistakeholder acceptance more quickly. MEND has essentially developed a toolbox of technologies available to all stakeholders, including operators, regulators, and consulting engineers.

Through MEND, Canadian mining companies and provincial and federal departments have reduced the liability due to acidic drainage by at least $400 million. This is an impressive return on an investment of $17.5 million over eight years. A decision has recently been taken to extend the program by three years. MEND 2000 will focus on verification and monitoring of large-scale field tests of MEND-developed technology as well as information exchange within Canada and with other countries.

Mining in Sensitive Ecosystems

Mining in sensitive ecosystems presents its own unique set of challenges. Not only are these areas often important reservoirs of biodiversity, but they are also often home to indigenous peoples, some of whom may not have had much contact with modern society. This may particularly be the case in tropical regions. Attention to this issue has been sparked by the Omai incident (pollution caused by breakage of a tailings dam) in Guyana and by pollution problems associated with the use of mercury by artisanal gold miners in the Amazon basin.

Corporate commitment to best practices and consultation with all stakeholders throughout all stages of exploration, development, production, and mine closure are integral factors for success. These challenges, while significant, can be met through the application of environmentally sound technologies and environmental management systems. The impact on the environment and on local people can be kept to a minimum so that the outcome of the mining project is advantageous to all parties involved. Examples illustrating progressive approaches to nonferrous mineral development follow.

The Herradura Project—Minera Penmont

Among the key policy objectives of Minera Penmont, a joint venture between Grupo Peñoles (Mexico) and Newmont Gold (USA), are to protect the health and safety of its employees, to minimize the adverse impacts of industrial and mining activities on the environment, and to contribute positively to the surrounding areas in which it operates.

The Herradura gold mining project located near the Mexico–United States border serves as a case in point. At an early stage of the project, Peñoles and Newmont recognized that some environmental impact would be unavoidable during the mine's development and operation, especially in the open pit area, the heapleach area, and the waste rock dumps. Consequently, during the planning and design stages, the company assigned the highest priority to the environment, particularly in the exploration, design, and development stages, when special attention was given to preserving the area's ecosystems. Moreover, the surrounding environment was carefully assessed with a view to facilitating restoration work during the operation and at the end of the mine's life.

The company worked closely with environmental authorities to establish a set of measures to be undertaken before construction began on the mining and mineral-processing facilities. The measures included the following:

- Transplanting over twelve thousand plants into two preestablished areas chosen for their similarity to the plants' original site, in order to ensure the survival of the unique flora in the area that would be impacted by the mining project;
- Preparing a special waterproof area to prevent soil contamination from the use of cyanide in the gold recovery process. The area consists of compacted clay bedding covered by a plastic lining and a layer of sand and crushed material; this design allows for the easy collection of dissolved metal-enriched solution;
- Building an additional enriched-solution collector vessel in case of emergency situations and/or higher flows resulting from severe rainfalls; the vessel is surrounded by a fence and is capped by a floating cover in order to keep out wildlife;
- Recycling all water used in the process, because water is a limited resource in the area; waste water is treated and then used to irrigate green spaces;
- Implementing a series of actions to prevent dust pollution, including the use of dust depressor liquids during the crushing process and during the transportation of mineral and waste rock, keeping the principal access roads damp;
- Choosing special equipment that will keep noise pollution to a minimum; and
- Using environmentally sound technology throughout all stages of the mining and milling process to ensure worker health and safety and to control atmospheric emissions to levels below national standards.

Minera Penmont is also committed to a permanent reforestation campaign on the twenty-six hundred hectares of the site and is promoting similar efforts in the local community. The company will also design and implement a program to protect the wildlife located on the site, in particular those species with special status under Mexican law.

Minera Penmont is confident that the Herradura project can be carried out in harmony with the local ecosystem and in a way that contributes to the economic and social growth of the local community without compromising the needs of future generations.

Minera Las Cristinas in Venezuela

Another project warranting attention is the Los Rojas development in Venezuela, implemented jointly by Minera Las Cristinas (70 percent Placer Dome and 30 percent CVG) and a local nonprofit organization, the Association Civil de Mineros Artesanales Los Rojas (ACMALR). The project involves the development of a small-scale, forty-tonne-per-day gold mining project within the Las Cristinas concession in southeastern Venezuela. Although small in scale, the Los Rojas project is seen as critical to the success of a larger $600 million gold/copper mine, the Las Cristinas Project, to be undertaken by Placer Dome and CVG.

Las Cristinas is an important gold mining region in Venezuela. At its high point the property accommodated more than ten thousand people, including miners, their families, service providers, and others. By 1984, unregulated artisanal mining activities had become so chaotic and problematic that the government moved to bring the situation under control. A plan was developed to regularize activity in the region, which included, among other things, reserving of certain areas for organized small-scale mining and promoting large-scale industrial extraction of resources wherever viable.

When Minera Las Cristinas initiated organized exploration of Las Cristinas, there were still close to four thousand people living and working on the concession. In preparing the concession, the government resettled approximately twenty-eight hundred of these people on two sites, five kilometers east of the concession. However, many of the people continued with unauthorized artisanal mining as a source of income, as neither alternative employment nor adequate social infrastructure were provided at the time of displacement.

The concept of Los Rojas evolved from the realization that the security of the Las Cristinas Project and economic opportunity for local people were linked issues. Based on independent social and economic baseline studies and motivated by a concern with escalating tensions within the communities and unauthorized mining activity on the property, Minera Las Cristinas took the initiative of implementing the Los Rojas project. Simply put, the project recognized the local community's inherent right to a more secure livelihood. Consequently, the company proposed a set of new investments to build cooperation with the local communities.

The technical objectives of the Los Rojas project were two-fold. First, the project sought the gradual elimination of traditional forms of mining over time by testing open pit mining with mechanized overburden removal but selective removal of the pay material by hand. Second, the project sought to define milling options that would allow for the elimination of the use of mercury in the field, while maintaining or improving gold recoveries and the income levels of the miners. In addition to the favorable economic and social impacts of the project, eliminating the use of mercury would result in significant environmental benefits. In order to help coordinate community inputs, local people established ACMALR, the purpose of which was to improve the quality of life of their community and families. Minera Las Cristinas donated funds to the association that enabled it to hire a full-time administrative secretary, set up an office, and initiate its first programs.

As expected, project implementation was not without problems. A major crisis in community confidence occurred in 1996, when the community called for greater emphasis on concrete activities in the field and less on organizational development. By the end of 1997, this crisis was overcome and a number of accomplishments had been realized, including infrastructure development, the opening of a field infirmary, the commissioning of a meeting and administration center, three supervised excavations, the completion of metallurgical test work and preliminary mill design, the installation of a provisional nonmercury mill facility, and a field trip to Bolivia with the association and technical support group representatives. The purpose of this last initiative was to expose key people in the project to gold recovery and mercury management techniques practiced in Bolivia. It resulted in the finalization of the design for the mercury-free gravity mill in Los Rojas.

Substantial progress has been made to date, and Los Rojas serves as an example of how an international mining company and two communities of historically independent, but now disenfranchised, artisanal miners have been able to cooperate in a meaningful manner.

While these examples are only a few of many that serve to illustrate industry progress in meeting the challenges posed by mining in sensitive ecosystems, it is also clear that the controversy surrounding this issue is far from being resolved. The issue of mining in tropical regions is becoming increasingly important for ICME member companies. It is hoped that through the exchange of experiences of different corporations, many of the outstanding issues will be identified and addressed so that the nonferrous mineral potential of these regions and all its attendant economic and social benefits can be realized, while any harmful effects on the environment are minimized.

Conclusion

A new corporate culture is taking hold in the international nonferrous metals industry. The industry is rising to the challenges created by globalization and changing public expectations on environmental and social issues. The nonferrous metals industry has demonstrated that the goals of wealth and income creation, social progress, and a healthy and safe environment can be compatible. In fact, they often go hand in hand, as profitable companies are more likely to maintain high social and environmental standards to protect their reputation with customers and investors, as well as with regulatory authorities and society at large. It is also clear that improved cooperation among governments, industry, and other outside interests can play an important role in fostering environmentally sustainable, economically sound, and socially desirable human progress and development.

In looking to the future, the best approach to addressing the various environmental and social challenges facing the industry today would seem to be a combination of voluntary industry initiatives and government regulation. This would build on the more constructive spirit of cooperation on environmental issues evident in recent years.

While companies have made great strides in improving environmental performance and in responding to changing societal expectations, keeping pace with these changing expectations continues to pose a great challenge to the industry. Transparent reporting of environmental performance, inclusive consultative processes, and regulatory and policy approaches that harness the creative energy of the industry collectively hold out great promise that this challenge can be met.

ICME's Environmental Charter

Product Stewardship Principles

- Develop or promote metal products, systems and technologies that minimize the risk of accidental or harmful discharges into the environment.
- Advance the understanding of the properties of metals and their effects on human health and the environment.
- Inform employees, customers and other relevant parties concerning metal-related health or environmental hazards and recommend improved risk management measures.
- Conduct or support research and promote the application of new technologies to further the safe use of metals.
- Encourage product design and uses that promote the recyclability and the recycling of metal products.
- Work with government agencies, downstream users and other in the development of sound, scientifically based legislation, regulation and product standards that protect and benefit employees, the community and the environment.

Environmental Stewardship Principles

- Meet all applicable environmental laws and regulations and, in jurisdictions where these are absent or inadequate, apply cost-effective management practices to advance environmental protection and to minimize environmental risks.
- Make environmental management a high corporate priority and the integration of environmental policies, programs and practices an essential element of management.
- Provide adequate resources, staff and requisite training so that employees at all levels are able to fulfill their environmental responsibilities.
- Review and take account of the environmental effects of each activity, whether exploration, mining or processing, and plan and conduct the design, development, operation, and closure of any facility in a manner that optimizes the economic use of resources while reducing adverse environmental effects.
- Employ risk management strategies in design, operation and decommissioning, including the handling and disposal of waste.
- Conduct regular environmental reviews or assessments and act on the results.
- Develop, maintain and test emergency procedures in conjunction with the provider of emergency services, relevant authorities and local communities.
- Work with governments and other relevant parties in developing scientifically sound, economic and equitable environmental standards and procedures, based on reliable and predictable criteria.
- Acknowledge that certain areas may have particular ecological or cultural values alongside development potential and, in such instances, consider these values along with the economic, social and other benefits resulting from development.
- Support research to expand scientific knowledge and develop improved technologies to protect the environment, promote the international transfer of technologies that mitigate adverse environmental effects, and use technologies and practices which take due account of local cultures and customs and economic and environmental needs.

Statement of Community Principles

- Respect the cultures, customs and values of individuals and groups whose livelihoods may be affected by exploration, mining and processing.

- Recognize local communities as stakeholders and engage with them in an effective process of consultation and communication.
- Contribute to and participate in the social, economic and institutional development of the communities where operations are located and mitigate adverse effects in these communities to the greatest practical extent.
- Respect the authority of national and regional governments and integrate activities with their development objectives.

Communication Principles

- Provide a free flow of information on international environmental and developmental issues affecting the industry.
- Listen and respond to public concerns about metals and the environment.
- Develop and implement programs that communicate the benefits of a balanced consideration of environmental, economic and social factors.
- Present products, processes or services as being environmentally sound only when supported by well-founded contemporary data.
- Ensure information provided is candid, accurate and based on sound, technical, economic and scientific data.

Australian Minerals Industry Code for Environmental Management

Statement of Principles

Signatories to the Code are committed to excellence in environmental management through:

Sustainable Development
Managing activities in a manner consistent with the principles of sustainable development such that economic, environmental and social considerations are integrated into decision making and management.

Environmentally Responsible Culture
Developing an environmentally responsible culture by demonstrating management commitment, implementing management systems, and providing the time and resources to educate and train employees and contractors.

Community Partnership
Consulting the community on its concerns, aspirations and values regarding the development and operational aspects of mineral projects, recognizing that there are links between environmental, economic, social and cultural issues.

Risk Management
Applying risk management techniques on a site-specific basis to achieve desirable environmental outcomes.

Integrated Environmental Management
Recognizing environmental management as a corporate priority and integrating environmental management into all operations, from exploration, through design and construction, to mining, minerals processing, rehabilitation and decommissioning.

Performance Targets
Setting environmental performance targets not necessarily limited to legislation, license and permit requirements.

Continual Improvement
Implementing management strategies to meet current and anticipated performance standards and regularly reviewing objectives in the light of changing needs and expectations.

Rehabilitation and Decommissioning
Ensuring decommissioned sites are rehabilitated and left in a safe and stable condition, after taking into account beneficial uses of the site and surrounding land.

Reporting
Demonstrating commitment to the Code's principles by reporting the company's implementation of the Code and environmental performance to governments and the community and within the company.

13 Evolution of Environmental Practice during Exploration at the Camp Caiman Gold Project in French Guiana

Frederick T. Graybeal

The Camp Caiman gold project is on the south side of the Kaw Mountains, fifty kilometers southeast of Cayenne, French Guiana. French Guiana is an overseas department of France located about five hundred kilometers north of the mouth of the Amazon River. The project is accessible by paved road to within three kilometers of the exploration camp, then by forest road requiring four-wheel drive. The climate is typical of tropical rain forests with an average annual rainfall of four meters and an average maximum daily temperature of 32 degrees Centigrade.

In 1994 ASARCO Incorporated (Asarco), a large international mining company, identified French Guiana as a region with high potential for the discovery of gold. The Camp Caiman area had been identified previously by the Bureau de Recherches Geologiques et Minieres, a French government geological group, as a large area where gold values in soil were enormously high. In addition, several abandoned campsites and small ditches indicated the former presence of small-scale alluvial gold mining. Two exploration permits covering fifty square kilometers over the gold anomaly were acquired by Asarco in 1995 from the local French administration following a public tender. Three additional permits covering twenty-seven square kilometers were acquired in 1997, and an application for a further permit is pending.

Host rocks for gold mineralization are fine-grained sedimentary rocks which have been deformed along a zone of strong faulting (Adam et al. 1998). Gold mineralization is associated with zones of quartz veins and small amounts of sulfide minerals. All rocks have

Figure 13.1 French Guiana.

been intensely weathered to an iron-stained saprolite containing clay, sericite, and quartz to depths of 30–110 meters. By the end of 1997, Asarco had drilled 224 exploration holes using both a diamond drill and a small reverse circulation drill. The current resource estimate is 16.9 million tonnes, averaging 2.73 grams of gold per tonne, containing about 1.49 million ounces of gold. Roughly 15 percent of the gold anomaly has been explored to date.

The five Camp Caiman exploration permits are adjacent to the Kaw Biotope zone. This is an area of both wetland, which is a principal habitat of the Black Caiman alligator, and adjacent hills, where caves in resistant duricrust layers high along the north side of the crest of the Kaw Mountains are home to the elusive bird known as the cock-of-the-rock. A recent extension of this zone, and its classification as a natural reserve, reaches the southern edge of the five permits. The Kaw Biotope zone is largely coincident with the Kaw Wetland System, which is a Ramsar site. Ramsar sites are about 770 wetlands worldwide that have been identified for specific protective measures in a 1971 convention signed by ninety nations.

The crest of the Kaw Mountains, three hundred meters high, is only fifteen kilometers from the Atlantic Ocean. Onshore humid winds rise quickly upon reaching this ridge, leading to conditions typical of cloud forests, which are unusual in French Guiana. The combination of high rainfall and cloud forest environment favors high biodiversity and abundance of all species in this area. The flora of the Kaw Mountains have been grouped into five different landscape units (Granville et al. 1997), based largely on elevation. The majority of the fauna are presently thought to be relatively uniformly distributed throughout the area, with the exception of the two species just noted and in areas where hunting pressure has disrupted normal species distributions.

This chapter is a brief case history of how a large mining company approached and revised its environmental practice during exploration at the Camp Caiman gold prospect. Subjects discussed include initial exploration work, formation and functioning of an independent environmental oversight group, public communication efforts, and Asarco's response to the group's suggestions.

The Situation in 1996

Initial geologic mapping of the Camp Caiman gold anomaly was followed in mid-1996 by a trenching and modest drilling program that encountered potentially economic gold values in bedrock. Using a bulldozer, a campsite was cleared near a small stream, and roads were built for drill access. Use of a bulldozer to clear trees and underbrush for the campsite resulted in removal of the forest litter and soil layers, which were generally less than six inches thick. Rainwater runoff from the camp and silting of the adjacent stream was an immediate result, due in part to the absence of forest litter to filter the silt.

Where possible, drill roads were located to avoid all larger trees. This effort to preserve the forest canopy meant that the roads were shaded from the sun and, as a result, were constantly muddy. The muddy conditions led to several reroutings of access roads from the camp to the drill sites and, of course, runoff from the roads led to silting where roads crossed small creeks. Silting from road drainage was not generally serious because litter on the forest floor functioned as an effective filter. Road building was frequently a problem on steeper slopes because bulldozer operators found small creeks to be an easy place to deposit dirt and brush excavated for the road, which led to additional silting.

Contrary to popular misconceptions as to how exploration and mining groups approach their work, geologists worldwide are quite sensitive to the natural environment. Most entered the natural resource business because of a fondness for outdoor work and are very aware of impacts on the environment caused by their work. Asarco's exploration staff at Camp Caiman recognized and took steps to reduce the silting while preserving as much of the forest as possible. Exploration trenches were filled and recontoured after mapping. The campsite was also recontoured so that some of the rainfall runoff was diverted to the adjacent forest floor before entering the stream. Small oil spills in camp where machinery was serviced or repaired were also recognized as a problem, and concrete pads were built to avoid any further contamination. Trash that was not biodegradable was hauled back to Cayenne, and a waste-water treatment facility was installed. Finally, all operations were located so that streams draining the camp and areas of initial drilling flowed away from, and not into, the Kaw Biotope zone.

The constant dilemma was how to limit environmental impact, maintain reasonable limits on expenditures on a project where the economic potential was still unclear, and still keep the project advancing rapidly. There were no operating manuals for gold exploration in the rain forest and, since modern exploration work was new to French Guiana, there were few people with relevant experience. Frequent meetings with French administrators from the mineral, forest, and environmental agencies, including visits to the project, led to improvements in exploration practice. Although French regulatory agencies were helpful and have full authority over all project permitting, it still seemed more could be done.

Formation of an Environmental Advisory Committee

By 1996, worldwide public awareness of issues such as loss of tropical rain forests, loss of biodiversity, and sustainable development had risen to high levels, with the media reporting intensively on these subjects. As an extractive resource company, Asarco was acutely aware of these issues, and the often negative image of the mining industry portrayed by various nongovernmental organizations (NGOs) was of great concern.

By late 1996, drilling at Camp Caiman had revealed gold mineralization of sufficient grade and volume that budgets were being increased and frequent reports were being presented to Asarco's senior executives and board of directors. At that time Richard de J. Osborne, chairman of the board of directors and chief executive officer of Asarco, being aware of issues related to tropical rain forests and of Asarco's relative inexperience in these environments, instructed his management committee to investigate the formation of a group of scientific experts that would provide advice on our work practices. The need for a proactive approach was clear.

Initial visits were made by three senior executives and two senior members of the exploration department of Asarco to Dr. Thomas Lovejoy, a leader in the conservation of tropical environments at the Smithsonian Institution, and to Conservation International (CI), a well-regarded NGO. Dr. Lovejoy, we learned, had served on similar committees elsewhere in the Amazon River region, and CI, having recently published a report on best practices for oil exploration in the tropics (Rosenfeld et al. 1997), had become interested in mining issues. The work of both Dr. Lovejoy and CI was well known to members of Asarco's board of directors, who gave their full support to the idea of a scientific oversight group and to the plans of the company's management to make use of the advice from such a group to improve the work at Camp Caiman.

Following these initial visits, the advisory group was designed as follows:

- It would be a five-member group initially, to be called the Environmental Advisory Committee.
- Members would be internationally known biologists, botanists, and other specialists with expertise in tropical rain forests and in the social issues raised by mineral development.
- The Committee would advise Asarco on the environmental and social aspects of mineral exploration, mine planning, mining operations, reclamation, monitoring, and other subjects thought important by Committee members.
- The Committee would operate as an independent group, meaning they would not be employees of, or consultants to, Asarco and would receive no compensation for their work. All travel and meeting expenses would be covered by Asarco. Should the Committee members decide to speak out publicly, individually, or as a group, on aspects of the Camp Caiman or other Asarco exploration projects in French Guiana, they would be free to do so. Asarco would have sole responsibility for its actions and final decisionmaking authority; in the event of disagreement between Asarco and Committee members, Asarco would make its reasons clear for any decisions, and every effort would be made to reach a compromise.
- The Committee was to be part of the planning process, not simply a review group once plans had been formulated (and became difficult to change).
- Asarco employees would function as facilitators for Committee meetings but would not be formal members. A search on three continents was successful, with one member from each of the following organizations: Smithsonian Institution, Washington, D.C.; CI, Washington, D.C.; ORSTOM Centre de Cayenne, French Guiana; ORSTOM, Toulouse, France; and Centre National de Recherche Scientifique/Museum National d'Histoire Naturelle, Brunoy, France.

There was full awareness among the members of the new Committee of the potential environmental impacts of an open pit gold mine in the French Guiana rain forest. There was also a clear willingness to approach this initiative with open minds to see whether exploration practice might be improved and whether a more environmentally friendly mine might result, in the event that exploration work was successful. Reactions to the advisory committee concept from those outside Asarco were varied, as might be expected, including, "Why bother?" "It sounds crazy!" "A very dangerous concept." "Has it been done before?" "An important new idea." and "I am going to require it for all future projects."

Initial Meetings of the Committee

The first meeting of the Committee was held June 16–17, 1997; one day was spent at Camp Caiman and a day and a half in Cayenne at Asarco's office. A meeting was also held at the offices of the government's Department of the Environment without Asarco employees present. Discussions were wide ranging, but they focused on four fundamental issues.

First, in addition to the exploration permits just mentioned, Asarco had applied for an additional permit farther east at about the same time the extension to the Kaw Biotope zone was proposed. This eastern permit was located along a weak but persistent eastward extension of the gold anomaly and, as it turned out, was entirely inside the proposed extension of the Biotope zone. Committee members were firm and united in their opinion that Asarco should withdraw the application for that permit. Asarco had previously debated the wisdom of this application, but because the gold anomaly extended to the east, because good prospects are hard to find and this had become a very attractive project, because other exploration groups were applying for permits in the same area, and because the proposed extension to the Biotope zone had not yet been approved, the permit application was submitted. Following the meeting of the Committee, the permit application was formally withdrawn. There was concern in various quarters that this withdrawal would set a bad precedent, despite the fact that it was consistent with the environmental stewardship principles of the International Council on Metals and the Environment (ICME), of which Asarco is a founding member. Asarco then advised various NGOs and the media that the permit application had been withdrawn, mentioning the role of the Committee. This action is believed to have established that Asarco was willing to be flexible and responsive to recommendations of the advisory group.

The second issue was our exploration practice at Camp Caiman, where discussions focused principally on silting. The Committee advised that future clearing should be by cutting trees rather than bulldozing, that more elaborate stream crossings were needed, and that diamond drill sites should be reduced in size. Drill sites were large because after towing the drill and support platforms to the site, there had to be enough room to allow the bulldozer an exit route and also because drillers complained of the danger of being hit by large falling limbs. Following the meeting, an intensive program of revegetation of the campsite, bridge construction, and redesign of the principal access road was initiated at significant cost.

The Committee's concern about our exploration practice led to increased use of a small, rubber-tired, self-propelled combination diamond, reverse circulation, and auger drill (the Scout). The drill was a new design and had been difficult to use during initial trials in the wet conditions of the rain forest. However, the environmental advantage of this machine was that its small size and light weight allowed it to ride on the forest litter requiring only minor clearing of brush to gain access to drill sites. By comparison to the environmental impact of the road and site preparation required for standard diamond drilling, the impact of the Scout drill was small, and it became an important component of the close-spaced, shallow drilling required to establish continuity for ore reserve calculations.

The third issue was the design of an environmental baseline study. Again, there were conflicting concerns. The cost of beginning an expensive baseline study before defining a proven economically mineable reserve had to be weighed against the concern that the ecological equilibrium had already been shifted by the exploration work and that the baseline study should have started earlier. Asarco had gathered numerous proposals from scientists and consulting groups, but there was only limited experience in French Guiana regarding what might ultimately be needed. The Committee took an active role, advising on the experimental design of individual studies, coordination of the studies, and the management framework. The result was a much improved baseline study with an initial cost that seemed reasonable given the state of the project. Also, the Committee recommended that two separate studies be conducted, one to establish a baseline for the natural environment and one for the social environment, including public health and safety.

The fourth subject discussed was the need to hire an environmental manager for the French Guiana exploration program. Following a lengthy search, the position was filled with a French environmental engineer in February 1998.

Throughout the first meeting Asarco made available all information requested by the Committee, identifying those data it considered confidential. At the end of the first meeting the Committee acknowledged Asarco's willingness to be open and responsive to all subjects discussed. Members probably also noticed that we lacked operating expertise in tropical environments and saw that they could improve our exploration practice.The Committee also helped to facilitate an introduction with a local NGO that had been reluctant to meet with us. Committee members concluded the meeting with the comment that they would be willing to continue their participation as long as we continued to earn their trust. During the fall of 1997 a site visit was arranged for one Committee member who could not attend the June meeting. At this time a request was made to meet with at least one group representing local interests at future Committee meetings.

The second meeting of the Committee was held February 22–24, 1998, to review progress on the baseline studies, discuss possible metallurgical flow sheets and development alternatives for an open pit gold mine, and visit the project. As part of the baseline review, a citizen's group from a town of five hundred people near the project was invited to meet with the Committee. This group noted the potential benefits of a mine development but emphasized their lack of understanding of technical projects and the need for continuous communication to avoid any misunderstanding that might jeopardize the project. They understood the need to protect the rain forest but were worried that environmental concerns might cause social issues to be overlooked.

This was the first meeting where possible mining flowsheets were reviewed and the use of cyanide discussed. Technical aspects of a possible gold mine that are hard to change, such as open pits, use of cyanide, and the need for tailings disposal areas were discussed, along with some of many aspects that are flexible, such as low impact access routes, plant locations, and separate islands of operations as opposed to a single cleared area containing everything. The ore reserve was still growing, and it was emphasized that Asarco wanted the Committee to participate in the entire planning process, not simply in a final review of a completed plan.

During the visit to Camp Caiman, the Committee acknowledged Asarco's efforts to reduce silting and commented favorably on the low impact operations of the Scout drill. Asarco has encouraged the drill contractor to continue working with the manufacturer of the Scout to increase its depth capability and to reduce further the need to use the diamond drill. Some diamond drill sites were still considered too large.

The second meeting concluded with the Committee encouraging Asarco to develop a manual describing best practices for mineral exploration in the rain forest. There was also some discussion as to whether scientists working on the baseline study could estimate the impact of mineral exploration on the pre-exploration baseline. Finally, the Committee advised Asarco of the importance of a synthesis of ecosystem dynamics to bring the disparate baseline studies together. This could involve remote sensing studies and Geographical Information System (GIS) software not presently available.

Communication

In addition to formal, governmentally mandated communication, it is extremely important to establish a real dialogue with, and to gain the support of, local communities (W. G. Dallas 1992). To date, Asarco has described exploration activities in local newspapers and magazines and on the radio, has sponsored television programs about gold mining, has met with a variety of citizen groups, and has led numerous field trips to the Camp Caiman project. The Committee will not become a substitute for face-to-face meetings between Asarco staff and citizens of French Guiana or anyone else concerned about issues of mining in a tropical rain forest, and the Committee has advised us they will be monitoring our communication efforts.The Committee has also recommended the addition of a sixth member who would reflect the interests of the local communities, and a diligent search for that person is in progress.

Robert J. Muth, vice president of Asarco, has noted that when it comes to communicating it is easy to suffer from good intentions; "No good deed ever goes unpunished." If you communicate early in a project, you may be asked questions you cannot answer and be accused of covering up problems. ("They know but they won't tell us.") If you wait until late in the evolution of a project to communicate when you have most of the answers, you are accused of withholding information. ("They knew all the time and should have told us earlier.") Asarco's public communication efforts have been designed to be as open, factual, and current as possible without making unwarranted promises. Nevertheless, we are well aware that the normal suspicion with which some NGOs view the extractive resource industry has been further heightened by incorrect or misleading information that can be distributed widely and rapidly on the Internet. One never is able to know the ultimate size of the Internet audience to which the incorrect information becomes available.

There is evidence that the existence of the Committee has improved Asarco's access to certain NGOs that were initially unwilling to discuss their concerns directly with us. This one very important result has convinced Asarco that the benefits of having the Committee, in terms of access to new environmental expertise and an improved public dialogue on project objectives, are our ongoing investment of time and effort in fully responding to its concerns.

Discussion

Most of us are concerned with the environment and want to minimize the environmental impacts of our chosen professions. It is the effective implementation of those concerns and desires that is changing at

Asarco as we learn from our own work and from the experience and the concerns of others.

Given limited exploration budgets and the slim likelihood that any single exploration project will succeed, it is difficult to justify full environmental and social baseline studies before each exploration project begins. However, to provide more guidance in the early stages of future exploration projects, Asarco has begun preparation of a manual of best practices that will describe methods to reduce further the environmental impacts of mineral exploration. The Asarco manual will be modeled after a similar volume prepared by CI (Rosenfeld et al. 1997) for oil exploration. Preparation of the manual is discussed at weekly staff meetings in Cayenne, and the entire staff is involved. Increased communication is another low-cost and high-benefit exercise. In the past, Asarco attempted to maintain a low profile in exploration work, to avoid scrutiny by the competition; in French Guiana, stakeholders learned early of our activities, and their concerns are being addressed before planning begins instead of after the fact.

Environmental issues receive an enormous amount of attention in Asarco, and a vice president of environmental operations reports directly to the president. However, 1998 was the first year in the company's one-hundred-year history that the exploration department has had a full-time environmental manager assigned to a single field project. These and other attempts to be responsive to the Committee's recommendations are a high priority. Members participate without compensation, and the principal incentive for their continued participation is to see improved exploration practice. As a further example of compromise, and following discussions of several alternatives, the Committee has recommended a specific approach to mine design that they feel will reduce impact and accelerate reclamation over more traditional approaches, and it is likely Asarco will follow this recommendation. Imagine the response thirty years ago if an independent group of rain forest scientists had recommended a major change in the development design of an open pit mine!

Asarco considers mining to be a principal component of the concept of sustainable development as this concept is defined in the Bruntland Report (United Nations 1987). The products of mining carry electricity, are required for the construction of cars and homes, enable us to communicate with anyone at any time, and contribute to the production of food and medicine necessary to sustain the needs of present generations. Mineable ore reserves are the rocks in the ground that contain these products, and these reserves are nonrenewable, but the metals recovered from the rock, things like copper, gold, and aluminum, have infinite lives. As an example, roughly 90 percent of all the copper ever mined is still in use and will continue to be used for centuries to come, meeting the needs of future generations of the Earth's citizens. Further, the footprint of a depleted reserve can be substantially reclaimed and, with some creative planning, whatever imprint remains may be utilized as a resource for future activities unrelated to the former mining operation (for example, see Harcourt-Richards et al. 1992).

In the scheme of worldwide land use, mining requires very little land. The total land area that has been used for all mining activities since the initial colonization of the United States 350 years ago is less than 1 percent of the nation's land mass, and much of that has been or will be reclaimed. Nevertheless, off-site contamination of the air, water, and other resources can result if care is not taken. Few would question the importance of the products of mining to present and future generations, but if exploration and mining are conducted irresponsibly, they will not survive as free enterprise activities. This case history of Asarco's exploration work at the Camp Caiman gold project is an example of how one company is attempting to define and continuously improve a practical and responsible set of best practices to guide its work.

EDITOR'S NOTE

Since the writing of this chapter, Asarco has been acquired as a wholly owned subsidiary of Grupo Mexico S.A. de C.V.

NOTE

Reviews were provided by R. J. Muth, G. D. Van Voorhis, D. A. Fournier, and R. Cramail of Asarco. Views expressed are exclusively those of Asarco and the author.

REFERENCES

Adam, A., J-P. Pallier, F. Lillie, D. Fournier, and K. Robo. 1998. *Exploration at the Camp Caiman Gold Project, French Guiana*: Pocket Program, annual meeting, Orlando, Florida. Abstract, p. 180.
Dallas, W. G. 1992. Planning to avoid environmental conflict: The Tara story. In *Minerals, Metals and the Environment*. Institution of Mining and Metallurgy, pp. 547–551.
Granville, J. J. de, G. Cremers, and M. Hoff. 1997. "Le project de Reserve Naturelle de Kaw: Diversité des biotopes et de la flore de la Montagne de Kaw." Unpublished. 7 pp.
Harcourt-Richards, K., A. McCracken, and N. Rigby. 1992. Integrated resource planning and utilization. In *Minerals, Metals, and the Environment*. Institution of Mining and Metallurgy, pp. 391–403.
Rosenfeld, A. B., D. L. Gordon, and M. Guerin-McManus. 1997. *Reinventing the Well: Approaches to Minimizing the Environmental and Social Impact of Oil Development in the Tropics*. Policy Paper. Conservation International, Washington, D.C.
United Nations. 1987. *Our Common Future*. Oxford University Press, New York.

14 Conservation and Concession Contracts

*Environmental Issues in Mineral
Extraction Agreements*

David N. Smith
Cyril F. Kormos

Foreign investment in mining and petroleum projects in developing countries has typically been carried out pursuant to special agreements negotiated between transnational resource extraction companies and host country governments. Authority for government officials to enter into such contractual arrangements has usually been set forth in general foreign investment legislation, or mining or petroleum legislation. While in earlier decades considerable discretion was given to government officials to negotiate ad hoc agreements, increasingly at least some of the terms of these agreements have been mandated in legislation or in "model" or standard contracts (Smith and Wells 1975).[1]

Contracts and legislation have, until very recently, said little about environmental and social issues. Environmental provisions in mining and petroleum legislation have often been limited in scope or have made reference to general environmental legislation, which typically did not address in detail environmental issues related specifically to mining and petroleum development. Even in the 1970s, when governments began to exert greater "sovereignty" over natural resource development, little attention was paid to environmental and social issues. As a result of an increased flow of petrodollars into banks and increased lending by banks to developing countries and state-owned resource companies, and as a result of increased government bargaining power, contractual arrangements were strengthened to provide greater control by governments. Exploration periods were reduced, relinquishment requirements were accelerated, "ring-fence" clauses were introduced, and taxes, royalties, and governments' share of equity were increased (Walde 1988).[2] But a corresponding tightening, expansion, and enforcement of environmental provisions did not occur.

There are several reasons for this. First, in the 1970s there was still a general lack of concern in developing country governments about the need for environmental protection in resource industries. In addition, governments themselves were in a dual role of tax collector or joint venture partner (or both), on the one hand, and regulator on the other. The role of regulator was often at odds with the role of tax collector or joint venture partner. The lower the costs of doing business for a mining or petroleum company, the lower the deductions, the higher the net profits subject to taxation and the higher the after-tax profits available for dividend sharing with the government.

Other factors were involved as well. In undertaking cost-benefit analyses of resource (and other investment projects), government economists typically regarded environmental costs as externalities that did not immediately affect the government's calculation of whether the project represented an economic and efficient use of the nation's resources. Environmental costs were, at best, a problem for future generations to worry about. In addition, governments did not have available information on the history and extent of environmental problems in other countries, both developing and industrialized. The legacy of polluted waterways, destroyed marine ecosystems, ghost towns, deforestation, and illnesses that characterized so many concession areas around the world simply was not part of the education of those charged with negotiating agreements or designing mineral and petroleum legislation. Even where some environmental standards existed, few government officials had the ability to determine the specific environmental protection practices required for particular projects. In addition, trained and well-equipped environmental enforcement personnel were lacking.

Local communities were also often unaware of existing or impending health and environmental problems and were focused, like governments, on economic issues of employment and revenue sharing. In those instances where some degree of awareness existed, local communities typically lacked the political power to influence the way in which the transnational company operated.

Finally, even transnational mining and petroleum companies that had good environmental records in their home or other countries were likely to relax standards in situations where environmental rules were vague, where there was no monitoring by a highly trained environmental administrative unit, and where project developers and managers in the field felt the need to respond to home office demands for cost control.

The 1980s witnessed a dramatic fall in the prices of many minerals and in the price of petroleum, as well as a dramatic decline in bank lending to developing countries and state-owned resource companies.

As a result, contracts were often renegotiated or redesigned to reflect the shift in bargaining power to transnational companies, as competition for investment in mining and petroleum increased. Exploration periods were extended, relinquishment requirements were relaxed, royalties and taxes were lowered, ring fence clauses were eliminated, and requirements for government ownership of equity were abolished. As governments began to compete more forcefully for foreign investment generally, and for foreign investment in mining and petroleum in particular, new foreign investment laws in developing and newly industrializing countries were introduced, and new mining and petroleum legislation followed. This legislation typically reflected a perceived weakening of the bargaining power of these countries vis-à-vis transnational companies.

The new mining legislation of the last decade typically provides for improved investment and approval procedures, security of title and land use, the right to assign mining rights, clearer rules for moving from the exploration stage to the mining stage, confidentiality of company data, the right to use mining rights as collateral, and the right to repatriate profits (World Bank 1996). Some legislation, such as that of Peru, Chile and Mongolia (see the 1997 Mongolia Minerals Law, article 20), provides for the negotiation of stabilization agreements that freeze terms in a contract for a period of time.

The Environmental Dimension in Mining and Petroleum Laws and Contracts

Despite the general liberalization of foreign investment legislation and mining and petroleum law, there has also been some increased attention to environmental and local community issues in mining and petroleum legislation and contracts. Much of this is still in its infancy and is inadequate in terms of providing detailed guidelines for companies, but legislation and contracts now typically provide for an environmental impact assessment (EIA) by companies. Some contracts and legislation go well beyond this.

The Philippines Mining Act of 1995, for example, includes detailed provisions for "environmental protection and enhancement" programs; environmental monitoring and auditing, contingent liability and rehabilitation committees, and decommissioning plans; a mine waste and tailings fee and a fee reserve fund; an interagency contingent liability and rehabilitation fund steering committee; and the Mine Environmental Protection and Enhancement Office. The Philippines legislation also provides for the involvement of representatives of local government units, nongovernmental organizations (NGOs), and community organizations on committees designed to evaluate environmental protection programs (Philippine Mining Act 1995, chapter 16).

The increased attention to environmental protection in the Philippines legislation, as in that of some other countries, stems from the increased political power of local communities and NGOs, environment-related law suits, increased awareness of environmental problems within government, and the country's own mining history. There is also more awareness of the history of mining in other countries, resulting from improved communications and better informed consultants.

Despite advances in some countries, mining and petroleum legislation is typically short on specific methodologies for dealing with environmental issues. The challenge today for developing and newly industrializing countries is to continue to open and liberalize markets so that they may compete effectively for a new wave of mineral and petroleum investment without doing so at the expense of the environment or local populations. A common response from environmentalists in addressing the development versus conservation issue is to prescribe more stringent regulatory frameworks. However, the task of achieving a balance between creating statutory controls providing comprehensive social and environmental protection and creating a climate that will promote growth and attract investment is extraordinarily difficult; to do so rapidly is even more so.

This is not to suggest that developing countries should not begin working toward this goal immediately, as the Philippines, for example, has done, but only that drafting appropriate environmental legislation and establishing the institutional capacity to implement it is often a lengthy, expensive, and difficult process (U.S. Office of Management and Budget 1997). Because designing effective, comprehensive, and project-appropriate regulations (environmental and otherwise) takes time and experience, contracts will continue to be a major vehicle for legislating rules relating to the protection of the environment and local communities. However, even bilateral contracts will not be enough to deal with these issues. Given the conflicting roles of government as tax collectors and equity holders on the one hand and regulators on the other, other parties, particularly NGOs and local communities, will need to play an increasingly active role in encouraging and ensuring attention to environmental and social issues.

Obstacles to Environmental and Local Community Protection in Resource Legislation and Contracts

Drafting an environmentally and socially beneficial contract is complicated by a number of factors that militate against effective protection of the environment and local communities, even where the need and relevant standards are recognized. This section summarizes these obstacles.

Transfer and Assignment

First, certain "liberalizing" provisions of recently adopted mining legislation are problematic. Much of the new legislation, for example, allows for transfers, assignments, and "farming" out to other companies, as well as for mortgaging of mining rights.[3] In some cases transfers can be made without government approval. Even where approval is required, it is often unclear whether the transferee's environmental track record is to be taken into account. In addition, in an attempt to reduce opportunities for corruption and delay, increased emphasis has been put on making the transition from exploration to mining stage as automatic as possible. But it is at this transition stage that the government should have the opportunity to evaluate feasibility studies and social and environmental impact assessments (World Bank 1996).

Fiscal Issues

Fiscal provisions governing mining projects may also work against environmental and social protection. High tax rates may lead to lower environmental diligence because they may lead to cost-cutting elsewhere. In addition, fixed royalties and tax rates that are uncalibrated to the size and grades of ore bodies may discourage the mining or processing of lower grade ores (which may be more expensive to extract), leading to more waste material and more toxic leaching (Hettler and Lehman 1995).

Perceived Bargaining Power

There are other reasons why drafting environmentally and socially sound contracts is difficult. Many developing countries often operate under the assumption that they do not possess the kind of leverage necessary to insist on stringent social and environmental provisions. Convincing government negotiators that they do have the bargaining power to require a transnational corporation to hold to certain standards is not an easy task. Furthermore, although progress has been made in defining best practices in mining and petroleum development, it is not always clear what best practices should be in particular instances and how these should be incorporated into the contract (Rosenfeld et al. 1997). Negotiation of the Ok Tedi agreement in Papua New Guinea demonstrates how difficult it is to determine appropriate standards of operation for a particular project in a particular geographic setting (Pintz 1984; IUCN 1995).

Financing, Participation, and Monitoring

Three additional issues have a significant impact on the effectiveness of environmental and community protection: financing, inadequate participation of local communities, and the lack of institutional mechanisms to monitor and enforce compliance.

Ensuring Financing for Environmental Protection

Resistance on the part of the private sector to spending on environmental protection is easily explainable. Until recently there have been no real costs that might appear on a transnational corporation's balance sheet as a result of failing to protect the environment. In addition, the prospect of lawsuits or penalties for environmental damage that would cause companies to internalize these costs seemed nonexistent or remote.

Why governments have not insisted that companies set aside money to preserve the ecological and social integrity of project sites is a more complex question. One reason is that many countries have lacked the technical expertise to determine precisely what prevention, mitigation, and remediation measures are necessary for projects. Another reason, mentioned earlier, is the dual role of the government as tax collector and dividend receiver on the one hand and environmental regulator on the other. In addition, countries anxious for hard currency and technological expertise have often not felt they are in a position to ask for expenditures on environmental protection. Transnational corporations frequently characterize projects as "marginal," thereby emphasizing the need for keeping costs low.

The relationship between host government and transnational corporation can be even more complex. Although many companies still argue that their duty to society ends when they pay their fair share of taxes and meet their fiduciary obligations to their shareholders, generally speaking, the corporate world is more than aware of environmental and social obligations. U.S. companies operating within the United States, for example, are also increasingly being held liable, in many cases strictly liable, for the social costs of their actions, and shareholders, consumers, and the general public expect more from the corporate world than ever before. In developing countries, however, it may be politically difficult for a corporation to be proactive on social and environmental issues, especially if the corporation is foreign. A company's well-intentioned actions may be viewed as a threat to governmental sovereignty. And governments may simply not be interested in addressing environmental issues or the concerns of local communities (Walde 1992).[4]

Some of these obstacles to effective environmental regulation are being removed gradually. As described earlier, the investment climate in developing countries is becoming more attractive as developing countries liberalize and become more democratic. The move away from exchange controls, combined with increasingly transparent administrative regimes and security of tenure and stabilization clauses, may make transnational corporations more amenable to accepting expenditures on environmental protection. The fact that natural resources in developing countries are in increasing demand because mineral reserves in developed countries are dwindling is an equally important factor that makes it easier to require more of transnationals. As developing countries become more adept at collecting their fair share of revenue through better design of fiscal regimes and better tax administration, environmental protection costs may seem less threatening. Just as important, local and international NGOs are playing a growing role in pressuring companies and governments into environmental protection spending.

A number of financial mechanisms have emerged in mining and petroleum legislation and contracts that can be used to ensure that companies provide adequate funds to address environmental and social concerns. These mechanisms should be used to provide funds for activities beyond the preparation of EIAs and social impact assessments (SIAs) and environmental management plans (Gao 1994). These assessments and plans should be financed by the company involved and designed by independent contractors, because both parties must have reliable information concerning the risks posed by the project, the options available, and the measures that must be taken to avoid or minimize those risks. These measures should therefore be presented as baseline costs rather than additional environmental expenditures.

In addition to paying for these baseline costs, a company should also present evidence of an environmental insurance policy and post an environmental performance bond prior to beginning operations. These are necessary precautionary measures, regardless of the financial health of the investing company or the investor's parent company at the outset of the project. Many companies that have been seriously out of compliance with environmental provisions or regulations are companies that were apparently financially sound when they began a project but later experienced difficulties and sought to reduce expenditures by decreasing environmental monitoring. Insurance coverage should therefore cover general liability as well as specific environmental and social liabilities arising from the specific activities undertaken in the project (Rosenfeld et al. 1997; Freeman 1997).

Because insurance companies rarely disburse funds quickly, and because large claims are often subject to litigation concerning the scope of the policy, a performance bond is a necessary additional security. A performance bond serves two functions. The first is that it allows for immediate disbursement if certain contingencies occur. The second is that it provides an incentive to the company to prevent environmental damage: If the environmental damage does not occur, the company can retire the bond at the conclusion of the project. In such a case, the only money lost is the interest the company might have paid to a third party surety to guarantee the amount of the bond (Bacow and Wheeler 1991).

Funds from insurance coverage and performance bonds by definition become available only after problems have occurred and have been detected. However, financing may also become an issue during the course of normal operations. If a project is less profitable than predicted, the company may find that it has difficulty meeting its baseline environmental obligations. Additional financing might also be needed in the event that problems occur that were not anticipated in the environmental management plans, or problems that prove greater than anticipated (Minkow and Dunning 1992; IUCN 1995; Pinz 1984). Finally, financing may be necessary to support a long-term contribution to biodiversity conservation efforts as well as for environmental mitigation and remediation in the project area.

In many instances, trust funds have been established to meet the costs just described. A trust fund performs the dual function of allowing the company to expand environmental expenditures required under the environmental management plan and providing a mechanism for assisting in biodiversity conservation long after the project has ended. The process of establishing the fund, its bylaws, and its board of trustees can be complex. It may be that the company must first assist the country in establishing a national fund and then set up a subaccount for its own project. It may also be that the company will be unwilling to endow the entire fund from the outset and that a schedule of contributions, combined with proceeds from a green tax, will be necessary to maintain the fund at acceptable levels (Rosenfeld et al. 1997).

How the fund's assets are disbursed is also an important issue. If the project site is in a remote area, it may be reasonable to spend the interest and invade capital, allowing the fund to sink a certain number of years after the project ends. If the project is in a populated area, it might be preferable to maintain a certain amount of capital to ensure a revenue flow for local communities. Finally, it is important that the fund assets be managed independently, rather than combined with other government money, to ensure that the fund is not used for purposes it was not intended to cover.

Participation of Stakeholders

Another factor limiting implementation of environmental and social safeguards is lack of participation by affected parties. Much has been written about the critical need to include local parties in planning and decisionmaking in order to ensure that they are compensated, their cultural integrity is respected, and their human rights are not violated (White 1994). Unfortunately, the literature frequently does not do justice to this task, one that is as complex as it is important.

To include stakeholders in a planning process first requires identification of stakeholders. In some cases, identifying who will genuinely be affected by a project, and who is entitled to a share of revenues from the land being developed, is not at all simple. There have been several notorious incidents involving NGOs attempting in good faith to represent indigenous or local interests but failing to include all affected parties. Land tenure and demographic studies may be necessary to settle territorial disputes or to detect migrant populations.

It may also be difficult to identify those individuals who have the authority and the competence to represent the various stakeholders and to negotiate on their behalf (White 1995). To complicate matters, those stakeholders who are identified may well have strongly competing interests, which becomes a serious problem if certain groups feel that they have received inferior compensation packages. In addition, it can be difficult to convey in understandable terms the scale and potential environmental effects of a large-scale mining project to stakeholders, especially if the stakeholder community is eager for employment and a source of revenue.

Legislation and contractual provisions can prescribe systematic SIA, conducted by independent anthropologists and NGOs, to study relationships among stakeholders and to conduct land tenure and demographic studies prior to exploration, prospecting, or development. Contracts must also include compensation clauses in the form of trust funds and royalties, as well as explicit recognition of rights arising under national law and international human rights instruments.

One of the factors determining the outcome of any negotiated agreement is the structure of the negotiation—that is, the number of parties involved and how they negotiate. Where indigenous groups are concerned, questions of sovereignty may arise, and the national government may not allow separate negotiations with indigenous groups. If indigenous groups do negotiate with a transnational company independently, it may be necessary to conduct a second negotiation between the government and the indigenous groups. Or it may be that a single complex negotiation involving the three parties will be held. The cost of a protracted negotiation and the skill of the parties also play a role in determining the final form of an agreement and the degree of protection afforded to the environment and local communities. Although in some cases it may not be possible to overcome governmental reluctance to allow direct negotiation by transnational corporations or NGOs with local communities, agreements must nevertheless ensure that, at a minimum, information about affected parties is available and that indigenous and local communities have some form of input and representation (White 1995).

Monitoring

Another major factor limiting successful implementation of environmental and social safeguards is lack of monitoring. Even well-intentioned companies have the tendency to relax their environmental standards when operating in remote areas for long periods of time. Well-intentioned companies may also fail to take note of certain environmental and social impacts of their actions because they lack proper mechanisms in the home office for gathering information concerning their project or simply because project managers are less concerned about environmental and social issues than the senior managers who sign the contracts. Perhaps more to the point, the prospect of any company monitoring itself objectively, even assuming the necessary expertise, is probably illusory.[5]

It may be similarly unrealistic to expect a government that lacks funds for basic needs to establish the institutional framework necessary to monitor resource extraction projects. In addition, in countries that do have governmental agencies with the capacity to perform monitoring operations, those agencies may be the same ones that have the responsibility of stimulating investment in mineral and petroleum extraction. They may therefore be subject to the same biases as the companies themselves.

In some cases independent monitoring may be necessary, and a mechanism needs to be established to accomplish it. A first component is an environmental audit performed by an independent contractor and paid for by the company (Gao 1994). The independent contractor's primary duty is to monitor technical compliance with the environmental management plan. The second component is NGO monitoring. The function of the NGO is to ensure that local populations are informed of project activities and that they have a forum in which to lodge complaints regarding the project.

NGO participation is often a good idea regardless of whether the government has its own monitoring capability (assuming that a government is serious about environmental and local community protection) because NGOs have the ability to mobilize public opinion both nationally and internationally. NGOs also have the ability to put together the resources for litigation. The lawsuit filed against Texaco in the federal district court in New York involves a complaint of environmental damage caused by Texaco over many years of operations in Ecuador.[6] The case was brought by NGOs on behalf of indigenous groups in the Oriente region in Ecuador and seeks relief under the Alien Tort Claims Act. The case has generated substantial publicity, and the government of Ecuador (at first opposed to the litigation) has now joined as a party. A similar case was brought in Australian courts against an Australian mining company by indigenous groups from Papua New Guinea in connection with the Ok Tedi project (Hettler and Lehman 1995).

Formalizing NGO involvement could conceivably reduce an NGOs independence and dilute its effectiveness. It may be that the best mechanism would be to earmark funds for NGO assistance within a trust fund. However, including a contractual provision that, at a minimum, calls for assembling a team of private and governmental auditors at the beginning of a project (as in the Philippine mining legislation) will provide a clear guideline for the company and for affected parties. This provision, like EIAs and SIAs and environmental management plans, should be a condition precedent to exploration or prospecting activities.

Conclusion

The most encouraging development over the last several years in terms of environmental and biodiversity protection in mining and petroleum projects has been the increased role of local communities and NGOs in bringing environmental problems to the attention of governments, transnational companies, and the world. International conventions have also helped in creating a greater awareness of environmental issues generally, but governments and transnational mining companies (and domestic miners and domestic mining and petroleum companies) are unlikely to move significantly in the direction of greater environmental protection without political or legal pressure and clear, specific environmental plans of action.

The ultimate solution to bringing environmental protection to the forefront of contract negotiations involves a complex interplay of law, administration, and negotiation. Each case will therefore entail particular subtleties. However, several measures can facilitate the negotiation process and create a framework in which the likelihood for environmental and social protection will be higher.

As a starting point, governments need clear and specific legislative commitments to environmental protection in mining and petroleum legislation and the development of an environmental protection ethos within mining and petroleum ministries and other agencies of the government responsible for resource projects. In addition, there needs to be a clear legislative and administrative mandate that will lead to the selection of mining and petroleum companies that have well-established environmental protection track records and the capability to finance environmental protection.

Against this background, legislation and contracts need to establish fiscal regimes that encourage, and do not discourage, environmentally sound mining and drilling practices and that provide for bidding procedures that include environmental protection criteria. In addition there must be environmental protection provisions that specify best practices for specific projects, effective environmental protection monitoring, and investment in training of environmental protection personnel. Also needed are effective penalties for violation of environmental protection provisions and provisions for the involvement at the initial stage and later of local communities that have a stake in the protection of the environment.

Finally, NGOs and international organizations need to play a stronger role in facilitating the sharing of experience among countries with regard to environmental and biodiversity protection and in fostering greater interchange among resource companies (and mining and petroleum associations) and governments, local communities, and NGOs about environment, biodiversity, and local community concerns.

NOTES

1. See Indonesia (1997).
2. "Ring fence" clauses prevented a petroleum company from writing off losses in one exploration field against income in a second field.
3. See, for example, the 1997 Mongolian Minerals Law.
4. The Nigerian government's steadfast resistance to pressure from environmental NGOs with regard to oil development in eastern Nigeria (and the execution of Ken Saro-Wiwa and other Nigerian activists) and the government of Ecuador's initial objections to a suit brought by NGOs against Texaco for environmental damage allegedly caused by Texaco in Ecuador's Oriente region are two examples of government intransigence in the face of pressure from public interest organizations. See Sachs (1996) and Schemo (1998). See also Papua New Guinea (1996).
5. For an example of a company's experience with monitoring itself, see the experience of Union Carbide in Bhopal, India, in Morehouse (1994).
6. *Aguinda v. Texaco*, no. 93, Civ 7527 (S.D.N.Y. April 11, 1994). See Schemo (1998).

REFERENCES

Anaya, S. James. 1996. *Indigenous Peoples in International Law*. Oxford University Press, New York.

Bacow, Lawrence S., and Michael Wheeler. 1991. Binding parties to agreements in environmental disputes. *Villanova Environmental Law Journal*.

Freeman, Paul K. 1997. Environmental insurance as a policy enforcement tool in developing countries. *University of Pennsylvania Journal of International Economic Law* 18(2):477–486.

Gao, Zhiguo. 1994. Environment and petroleum agreements. *Journal of Energy and Natural Resources Law*.

Hettler, Jorg, and Bernd Lehmann. 1995. *Environmental Impact of Large-Scale Mining in Papua New Guinea: Mining Residue Disposal by the Ok Tedi Copper-Gold Mine*. United Nations Environment Programme, Nairobi, Kenya.

Hitchcock, Robert K. 1994. Endangered peoples: Indigenous rights and the environment. *Colorado Journal of International Environmental Law and Policy.*

Indonesia. 1997. Seventh Generation Contract of Work.

IUCN. 1995. *The Fly River Catchment, Papua New Guinea, A Regional Environmental Assessment,* International Union for the Conservation of Nature and Natural Resources Gland, Switzerland.

Minkow, David, and Colleen Murphy-Dunning. 1992. Assault on Papua New Guinea. *Multinational Monitor.*

Morehouse, W. 1994. Unfinished business: Bhopal ten years after. *Ecologist.*

Papua New Guinea. 1996. Prohibition of Judicial Proceedings Abroad (Mining and Petroleum Operations) Act (a law intended to nullify existing and future law suits in foreign courts with respect to environmental impacts of mining and petroleum operations in Papua New Guinea).

PETROPERU. 1991. *Peruvian Petroleum Legislation including Decree Law No. 22774, General Terms for Petroleum Contracts.*

Philippine Mining Act Revised Implementing Rules and Regulations. 1995.

Pintz, Sam. 1984. *Environmental Negotiations in the Development of the Ok Tedi Mine in Papua New Guinea.* World Resources Institute, Washington, D.C.

Rosenfeld, Amy, et al. 1997. *Reinventing the Well: Approaches to Minimizing the Environmental and Social Impact of Oil Development in the Tropics.* Policy Paper. Conservation International, Washington, D.C.

Sachs, Aaron. 1996. Dying for oil. *World Watch* 1 (May/June):10.

Schemo, Diana Jean. 1998. Ecuadorans want Texaco to clear toxic residue. *New York Times,* February 1.

Smith, David N., and Louis T. Wells. 1975. *Negotiating Third-World Mineral Agreements, Promises as Prologue.* Ballinger, Cambridge, MA.

U.S. Office of Management and Budget, Office of Information and Regulatory Affairs. *Report to Congress on the Costs and Benefits of Federal Regulations.* September 30.

Walde, Thomas. 1988. Investment policies in the international petroleum industry: Responses to the current crisis. In *Petroleum Investment Policies in Developing Countries,* edited by Nicky Beredjick and Thomas Walde. London: Graham & Trotman.

———. 1992. Environmental policies towards mining in developing countries. *Journal of Energy and Natural Resources Law* 10(4): 327–357.

White, Heather C. 1995. Including communities in the negotiation of mining agreements: The Ok Tedi example. *Transnational Lawyer,* University of the Pacific, McGeorge School of Law.

World Bank. 1996. *A Mining Strategy for Latin America and the Caribbean.* World Bank Technical Paper no. 345. Industry and Mining Division, Industry and Energy Department, World Bank, Washington, D.C.

PART V

Infrastructure for Sustainable
Development

15

Rethinking Infrastructure

Approaches to Managing Development on the National and Continental Scale to Reduce Conservation Impacts

Eliezer Batista da Silva
Gustavo A.B. da Fonseca
Amy B. Rosenfeld

Every major extractive industry, be it mining, oil development, or timber extraction, has one thing in common: The construction of major infrastructure, including roads, buildings, and energy sources, is always an important component of each project. Because this infrastructure can reach far beyond the boundaries of a specific project site, both in terms of distance and impact, proper siting and design is vital for ensuring biodiversity conservation. Even when not connected to a specific project, large-scale infrastructure development itself—particularly long-haul transportation, telecommunications, and energy projects—has the potential for extensive impacts on the surrounding environment, spanning across several cities, provinces, or even countries.

The creation and improvement of local, national, or regional infrastructural systems can promote economic development by facilitating the movement of people, goods, services, and ideas within nations, between neighboring countries, and to the rest of the world. At the same time, however, the negative environmental and social impacts of these projects can be equally far-reaching. Major transportation projects such as roads and railways cause direct impacts from construction, including siltation of streams and rivers, deforestation, air pollution, and bisection of parks, preserves, and habitats. These direct impacts are often followed by the potentially even more destructive indirect effects of both planned and spontaneous colonization, including land speculation, unsustainable agriculture, cattle ranching, and continued deforestation (Fearnside 1984). The same impacts can result when new rights-of-way are cleared for pipelines, power lines, and telecommunications networks. Increases in energy capacity through hydroelectric dams or coal-burning power plants can cause tremendous environmental degradation through flooding of large areas or high levels of air pollution.

The challenge for governments and project planners is to ensure—through careful planning and decisionmaking about siting and design—that these infrastructure systems are developed in a sustainable manner, minimizing environmental degradation while maximizing benefits to the affected populations. This chapter looks at how planners can increase the sustainability of planned infrastructure developments by setting regional and national priorities for biodiversity conservation and completing project-specific economic, environmental, and social impact assessments.

Regional Priorities for Conservation

From a conservation perspective, the first step in deciding where to site a major infrastructure project is deciding where not to site it. Every hemisphere, continent, or region has certain areas that rank higher than others in terms of biological diversity, and project planners should have a clear idea of these conservation priority areas before undertaking large-scale infrastructure development. Thus, before proceeding to a detailed analysis of a particular project, it is important to analyze the entire region using state-of-the-art mapping and imaging technology and economic, environmental, and social assessment tools.

Despite all the best intentions, it will be impossible to site infrastructure projects only in areas that are of little or no ecological priority or are already irreversibly degraded. Thus, we must focus our limited resources on the most ecologically important areas. The most commonly used criteria for the identification of priority areas are species richness, level of endemism (species that occur only in a particular region and nowhere else on Earth), community uniqueness, representation in protected areas systems, level of ecosystem integrity, intrinsic ecosystem and community resilience, and current threats. In addition to biological and environmental characteristics, it is also important to evaluate the main social and economic attributes of each region that is proposed for infrastructure projects.

Regional products and markets should be assessed to determine the potential economic costs and benefits of infrastructure development and the importance of the specific area for overall national or regional development. In addition, care should be taken to site infrastructure in a socially and ethically responsible way (for example, respecting the integrity of indigenous lands).

Specific Project Sustainability

Any assessment of the viability of individual infrastructure projects should look not just at economic and financial factors but also at potential environmental and social costs and benefits. The vast scale of many infrastructure projects means that conventional project assessment tools, such as environmental impact assessments (EIAs), may not be sufficient to fully understand these impacts. Traditionally, an EIA considers just direct, site-specific environmental impacts of a project over a limited period of time. To fully assess large-scale infrastructure projects, this process must be broadened, both spatially and temporally, and include social as well as environmental factors.

While a traditional EIA is geographically limited to a specific project site, an ecosystem-wide approach to impact assessment will look beyond project boundaries to judge both the direct and indirect effects of a project over the entire affected area. For instance, while a road project may have a relatively limited footprint in terms of direct land-clearing for the road's right of way, increased access to an undeveloped area in a region facing population pressures might lead to very extensive indirect land-clearing resulting from settlements and agricultural colonization along the roadway.

Brazil's Balbina Dam provides an example of a major infrastructure project that has had much wider ecological impacts than were originally expected or intended. The dam, which lies 150 kilometers north of the Brazilian city of Manaus on the Uatama River, was intended to provide 250 MW of power, about half of Manaus's electricity needs, and to speed development of Brazil's Amazonian wilderness (Powell 1989). However, by the time the first of the dam's turbines finally began operating in 1989, the $800 million project had flooded 2,360 square kilometers of forest, twice what was intended, to generate an average of only 100 MW (*Economist* 1992). The shortfall in power at Balbina and the need to flood a much larger area than was originally planned resulted from the fact that the surrounding land area was actually too flat to produce sufficient water pressure to run the dam (Simons 1987).

The Balbina Dam has had significant environmental and social impacts. The flooded region is fouled by the smell of hydrogen sulfide from decomposing vegetation, and heavy metals leaching into the water from the soil have killed fish and made the water undrinkable (Angelo 1995). Massive rescue efforts to save wild and threatened animals as the area was flooding involved 250 men per day and turned up hundreds of familiar and unfamiliar species, including some that were previously unknown to science (Simons 1988). Flooding has also displaced hundreds of people, including a third of the less than four hundred surviving members of the Waimiri-Atroari Indians (Powell 1989).

Expanding the scope of an assessment not just in space but also in time is equally important for fully understanding and mitigating the potential impacts of infrastructure development. The repercussions of a project must be considered in the long run, from planning through construction, implementation, maintenance, operation, and, if applicable, provisions for dismantling a project at the end of its lifetime. A persistent problem with infrastructure development is the lack of a long-term planning horizon, with projects thought of only in terms of the construction and implementation stages. An example of this problem is the common lack of provisions for long-term maintenance of a project. Recurrent costs for upkeep and maintenance are frequently not included in cost-benefit calculations or are given little weight because their effects are rarely readily apparent. Nevertheless, routine maintenance of roads and other facilities is an important long-term investment in the future, as the eventual costs of repair or replacement of a poorly maintained road can far exceed regular maintenance costs (Ostrom et al. 1993).

Finally, it is vital that this broad-based assessment take place before construction or even detailed planning begins—before there are vested interests and sunk costs that make a project that much harder to abandon. In this way, project planners can ensure that any impacts can be fully prevented or mitigated, since corrective measures after damage has already been done may be either too costly or too late. In some cases, the expected costs of environmental and social damage from a proposed infrastructure project will be much higher than any potential financial gain, thus suggesting abandonment of the project as it is currently conceived.

Economic Viability

While projects should be judged by traditional financial and economic considerations, additional criteria should be included to ensure the proper evaluation of the underlying economic assumptions of a project. Such evaluation may point to alternative approaches to infrastructure integration that reduce environmental impacts while maintaining economic viability. For example, the use of a waterway instead of a road might be a preferable and easily accomplished project adaptation to minimize environmental damage. Or, existing infrastructure, such as unpaved roads or out-of-service railroads, could be improved, repaired, or better maintained to achieve the same goals as a new development without requiring additional land-clearing.

The project should not only be financially viable from the perspective of direct project participants, it should also have a positive economic net present value, ensuring that its transportation or other benefits will exceed all costs. This is often a problem with projects that are based on unreasonably optimistic assumptions or have failed to account for all the costs of construction and long-term maintenance.

The economic viability of an infrastructure project is often directly related to the level of development or remoteness of its location. A project in a remote, less-populated region will have fewer beneficiaries and less opportunity to speed development of existing population centers. Generally, these are also the projects that stand to cause the most far-reaching environmental and social damage as well, by opening up access to undeveloped wilderness or facilitating contact with isolated indigenous groups.

Integration versus Penetration

Siting decisions can play an important role in assuring that a project achieves a maximum economic return on its investment while at the same time minimizing environmental and social costs. Large-scale infrastructure projects should integrate existing important and complementary economic centers of production, commerce and consumption rather than open up unsettled areas to development. In general it is usually prudent to avoid penetration of wilderness areas with new large-scale infrastructure because there is no other existing infrastructure nearby and because the wilderness areas are generally poorly understood from a scientific perspective.

The Transamazonia Highway is an example of a wilderness penetration project with many costs and few benefits. As originally proposed in 1970, the highway was to stretch for fifty-four hundred kilometers from the Atlantic coast of Brazil to the Peruvian border, passing for three thousand kilometers through the Amazon basin and bisecting national parks. At the time the highway was built, the government justified the development of the Amazon wilderness as necessary to alleviate the poverty and overpopulation plaguing the northeast after a drought. In addition, expected agricultural surpluses and new supplies of timber and mineral resources were intended to enrich the nation, and a large settlement of Brazilians in the Amazon was expected to fend off a feared influx of migrants from other nations hoping to settle the vast unpopulated space (Fearnside 1986).

Colonization, however, was not terribly successful or profitable along the twenty-three hundred kilometers of the highway that eventually were built. The soil on the newly cleared land was poor, leading to declining yields and increasing pests and disease after only a few years. Most farmers found it more profitable to sell their land to cattle ranchers or simply abandon it and clear another section of forest (Goodland 1990). The majority of settlement along the road was not government-planned, but rather spontaneous migration; in some areas of the newly opened wilderness, population growth was as high as six times the national average. A lack of services and land for this new wave of migrants led to increasing stress on forests and other biological resources (Fearnside 1986).

Systemic Development

In constructing infrastructure links between existing economic centers, combining several forms of infrastructure—for example, energy and telecommunications transmission lines, oil and gas pipelines, roadways, and railroads—in a single development corridor will help to minimize overall effects on the surrounding ecosystems and cultures and may save time and money in construction. In some cases, to maximize efficiency and economic and environmental benefits, planners should also look beyond local and national borders in designing these infrastructure belts. For instance, the closest source of power for an area or the closest major port or market might be across a provincial or national border.

Even if a project alternative that follows existing infrastructure corridors will have more direct costs, in many cases these financial costs can be justified and outweighed by the vast reduction in ancillary environmental and social costs as a result of the project. For example, in the Maya Biosphere Reserve in northern Guatemala, a region already under great pressure from colonization, a recent oil pipeline construction project highlights the potential impacts of failing to consider the indirect impacts of infrastructure development. In 1995, a small oil company operating within the Laguna del Tigre National Park in northeastern Guatemala began construction of a pipeline to transport crude from its oil field to a processing center. Although an oil road already provided a cleared right-of-way for access to the oil field, the pipeline, for financial reasons, was constructed along a shorter, more direct trajectory rather than along the path of the existing road. As a result the pipeline crosses significant stretches of primary tropical forest and has opened a new right-of-way into the national park.

Initial environmental assessment of this project did not consider alternative pipeline paths nor the potential indirect environmental impacts, such as colonization, of siting the pipeline away from the existing road. Nevertheless, even before the pipeline was constructed, there was already evidence of an increased influx of colonists and expansion of slash and burn agriculture along the existing oil road, and indications are that the same may occur along the pipeline right-of-way (Sader 1996).

Environmental Viability

In addition to conducting regional priority analyses and avoiding indirect, ecosystem-wide impacts, it is equally important to look at the specific, direct ecological impacts of each project. To minimize environmental impacts, projects should not take place in a high priority area for biodiversity conservation, as determined through the priority-setting methods discussed earlier, and projects should never be constructed within a recognized or proposed protected area.

After a siting decision has been made to reduce the probability of wide-ranging environmental and social impacts, design, construction, and implementation of a project should be completed with a minimum of environmental damage. All projects should be designed to ensure the sustainable use of natural resources, as well as the use of the best available pollution prevention and waste minimization technologies, during both construction and implementation. Plans should be made for complete and proper disposal of any generated wastes, and the project should provide for environmental mitigation, where appropriate. Such mitigation might include the creation of a well-enforced national park or protected reserve along a road, railway, or riverway that is at risk of leading to colonization in sensitive areas.

Social and Institutional Viability

Ensuring the social sustainability of a project includes monitoring the effects of the project on local people, in terms of health, knowledge, culture, and income. The project should not threaten the land rights or traditional ways of life of local indigenous people, and any negative impacts on isolated indigenous groups should be avoided. The project also should not displace members of any local populations against their will, as may happen with the removal of communities from areas flooded by hydroelectric projects. Local stakeholders should be fully informed of project plans in every phase, from design through implementation. To increase value to these local people, the project should help alleviate poverty in the affected region and, of course, should not lead to the impoverishment of any local peoples. Projects can contribute to the welfare of the local people through employment opportunities, expanding the market for local products, and increased access to economic centers and social services such as health care and education. Finally, the project should not impact or modify important archaeological or cultural sites.

The most profitable, ecologically sound, and socially sustainable project may fail, however, if the local, regional, or national governments and institutions in the project area are incapable of successfully guiding a project from planning through implementation. Thus, the appropriate policies, legislation, and regulations must be in place to facilitate the project, and there should be representative government approval and involvement at national and local levels. In addition to legislation, the appropriate local, regional, or national institutions should exist or be created to carry out proposed projects. While rudimentary institutional capabilities are a prerequisite for project success, any project that contributes to institution-building in the affected area gains additional appeal.

Integrated Decisionmaking

By overlaying the results of regional or national conservation priority setting with assessments of the economic, environmental and social viability of various proposed projects, planners can form a series of recommendations for how infrastructure development should proceed in a certain area. If this integrated analysis shows that a project will be economically viable and would have a minimum impact from an environmental standpoint on an area of low to moderate priority, planners should generally go ahead with the project as it is proposed. This would be the case if the area under direct influence from the project has already been altered to an irreversible degree, or if the project only tangentially affects a low or moderate priority region.

Determining a course of action becomes more difficult when the project will have relatively significant impacts on an area of moderate to high priority. Such projects should only proceed if site-specific or technical mitigation techniques may be used to bring foreseeable impacts down to acceptable levels. For example, a proposed highway might be rerouted to avoid bisecting an existing protected area, a particular fragile ecosystem, or indigenous lands. Or, instead of using dredging to expand a riverway, alternative technologies of barge construction may make dredging unnecessary.

Further economic and technical analysis may indicate that the need to redesign a project, and the additional investment required to make it environmentally sustainable, would make the project unprofitable. In this case, it may be possible to get funding from other sources to cover the additional costs of environmental protection or mitigation. If, however, a project will have a significant impact on a moderate to high priority area, and the additional costs are excessive and far outweigh the benefits, it should be abandoned.

Conclusion

Major infrastructure development can provide huge economic and social benefits to a country or region by increasing the quality and availability of goods and services flowing into and out of an area. However, if not properly planned, this development can also lead to equally large economic, environmental, and social costs, resulting from degradation of important biodiversity areas, population pressures, and failed agricultural development. By ensuring that projects are viable from both the economic as well as the environmental and social perspectives, project planners, whether they are from the public or private sector, can help ensure that infrastructure development proceeds in a sustainable manner, minimizing the costs to the environment while maximizing the benefits to the surrounding populations.

REFERENCES

Angelo, Mark. 1995. North American dam legacy a curse for developing nations. *Vancouver Sun*, February 20, p. A10.
Economist. 1992. The Beautiful and the Dammed. March 28.
Fearnside, Philip M. 1984. Roads in Rondonia: Highway construction and the farce of unprotected reserves in Brazil's Amazonian forest. *Environmental Conservation* 11(4):358–360.
————. 1986. *Human Carrying Capacity of the Brazilian Rainforest.* Columbia University Press, New York.
Goodland, R. 1990. "Tropical Moist Deforestation: Ethics and Solutions." Draft.
McNeely, J. A., K. R. Miller, W. V. Reid, R. A. Mittermeier, and T. B. Werner. 1990. *Conserving the World's Biodiversity* World Resources Institute, World Conservation Union, World Bank, World Wide Fund for Nature–U.S., and Conservation International, Washington, D.C., and Gland, Switzerland.
Osava, Mario. 1993. Brazil's Tucurui Dam yields energy and mosquitoes. Inter Press Service, May 18.
Ostram, E., Shroeder, L. & Wynne, S. 1993. *Institutional Incentives and Sustainable Development: Infrastructure Policies in Perspective.* Boulder, CO: Westview Press.
Powell, Stephen 1989. Controversial dam in the Brazilian Amazon starts operation. Reuters, February 18.
Sader, Steven A. 1996. *Forest Monitoring and Satellite Change Detection Analysis of the Maya Biosphere Reserve, Petén District, Guatemala.* Conservation International and the U.S. Agency for International Development, Washington, D.C.
Simons, Marlise. 1987. Dams threat to rain forest spurs quarrels in the Amazon. *New York Times*, September 6, p. 18.
————. 1988. On Amazon Noah's ark, all aboard (even wasps). *New York Times*, March 30, p. A4.

16 Environmental and Social Considerations in the Development of the Greater Mekong Subregion's Road Network

Robert J. Dobias
Kirk Talbott

In 1992, a program of cooperation was initiated among the six countries of the Greater Mekong Subregion (GMS): Cambodia, the Lao People's Democratic Republic (Lao PDR), Myanmar (also called Burma), Thailand, Vietnam, and Yunnan Province of the People's Republic of China (PRC). Referred to as the GMS Program, it aims to promote and support the double transition in the GMS countries from centrally planned to market economies and from agricultural to more diversified economies. The GMS Program is being supported by the Asian Development Bank (ADB), which houses the GMS secretariat, and by various bilateral and multilateral aid agencies.

Seven sectors have been targeted for cooperative activities, including the transport sector, which covers road, waterway, rail, and air transport. (The other sectors are energy, telecommunications, tourism, environment, human resource development, and trade and investment.) This chapter presents an overview of the environmental and socioeconomic setting in the GMS and summarizes the priority components of the GMS transport sector program as it relates to roads. We then discuss the environmental and social implications of road development generally, and the GMS road program specifically. We conclude by discussing ongoing and planned efforts to incorporate environmental and social concerns into road development in the GMS, and we offer ideas for improving the balance between road development and socioenvironmental priorities.

A Region of Rapid Growth and Rich Biodiversity

The ongoing currency crisis notwithstanding, the six countries of the GMS are in a period of rapidly accelerating economic growth and development, arguably greater than that of any other region on Earth. Historically separate river valley civilizations often in conflict with one another, the nations of the GMS are now increasingly bound by commonalities and regional bonds. They are also increasingly becoming an international arena for economic investment. Thailand's economy has been booming for nearly two decades, while Vietnam, Lao PDR, and Cambodia are beginning to emerge from years of war and isolation. Although the pace of development in Myanmar is less well documented, evidence suggests that it too is entering an era of economic expansion and rapid resource extraction (Brunner, Talbott, and Elkin 1998). Rapidly moving socioeconomic transformations in Yunnan Province, PRC, are destined to have substantial repercussions throughout the region.

This growth has been largely fueled by the conversion of natural capital (soil, forests, minerals, etc.) to financial capital. The countries of the GMS have an exceptional level of biological wealth. Although precise figures are difficult to obtain, Lao PDR, for example, maintains forest cover over approximately 40 percent of its land area, and until very recently about half of Cambodia was under forest. Yunnan Province, which, with an area of nearly 400,000 square kilometers and a population of forty million people is comparable in size to several Asian countries, has about one-quarter of its land under forest. The countries of the GMS are rich in species diversity. Cambodia has an estimated twenty-three hundred vascular plants. Lao PDR has more than 180 species of mammals and 600 species of birds. Cambodian inland waters harbor in excess of 200 fish species.

By promoting their comparative advantages in production costs and labor, resource-rich countries in the region have been able to establish for themselves lucrative export markets for natural resource-based products. In Lao PDR, for example, the export of energy from hydropower plants has generated approximately one-quarter of the country's annual foreign exchange revenues. Similarly, the export of forest products has been a significant source of national income, estimated at 54 percent of all foreign exchange earnings in 1991. Yunnan Province produces an estimated $300 million in forest products each year.

Economic growth has been accompanied by widespread environmental problems, some of which are beginning to undermine the very economic base that has made possible the substantial improvements in human welfare that are justifying the pursuit of rapid economic development. Forest degradation, led by unsustainable logging practices, has been a concern for all the countries of the GMS. Regional statistics (apart from Yunnan) indicate that forest loss may be on the order of 1 percent per annum, with approximately three-quarters of logging located in primary forests during the last decade. John Mac-Kinnon and others (1986) have estimated that the percent of loss of original wildlife habitat has been on the order of 70 percent in Lao PDR to 80 percent in Vietnam (MacKinnon et al. 1986). The average annual economic loss due to forest degradation in Lao PDR from 1982 to 1992 has been estimated at $87 million to $191 million, or 11–25 percent of gross domestic product (World Bank 1993). Conversely, the forests of Cambodia, if properly managed, could contribute up to $100 million to the national economy each year, a figure equivalent to more than half the entire 1994 national revenue (Talbott 1998).

Regional economic dynamics can exacerbate these environmental problems. Degradation of the natural resource base in Thailand, for example, is compelling producers, manufacturers, and swelling populations to look farther afield for supplies of marketable raw materials. As a result, Thailand is now expanding its supply lines into nearby countries where the price and availability of still untapped resources is most advantageous. Similar dynamics are coming to play in Yunnan Province.

Simply put, the nations of the GMS, including the comparatively sparsely populated and least developed countries of Lao PDR and Cambodia, are quickly reaching the limits of land and other natural resources in their heavily settled lowland domains. At the fringes of the lowland nation-states, frontier forests were typically relegated to the millions of indigenous "minorities," such as the Hmong, Karen, Akha, and Yao. In more recent years, however, they have also become home to millions of traditional lowlanders who have relocated in response to population pressures and marginalization dynamics.

Current and projected development patterns throw into doubt the long-term health and productivity of many of these areas. In the short run, those most likely to be adversely affected are the millions of rural inhabitants who rely to varying degrees on biological resources for sustenance or to supplement their incomes. Eighty-five percent of Laotians are employed in the agriculture sector, and 80 percent of energy consumption in Lao PDR is wood based. In 1990 to 1992, Cambodians earned $35 million from commercial inland fishing. An estimated 25 percent of Cambodia's twenty-three hundred species of vascular plants are used for medicinal purposes.

In addition, these resources provide a wide variety of ecological services vital to the health and productivity of the whole region. In the long run, therefore, the cost of the continued degradation of biological resources will inevitably extend to all the region's citizens. Given these dynamics, it is important that interregional road development plans give due consideration to the biological and cultural diversity and fragility of the GMS.

Major Road Projects in the GMS

Plans for the GMS transport sector are aimed at linking population centers and major tourist destinations; developing remote and low income areas; improving accessibility to markets; and reducing non-physical barriers to the movement of goods and people in the region. A study conducted in 1995 identified nine preferred road projects based on economic priorities.

Together, the nine projects account for over ten thousand kilometers. Five of the projects will involve Yunnan Province, four will affect Thailand, Lao PDR, and Vietnam; three Myanmar; and two Cambodia. While not all these projects may actually be implemented, and accurate figures for many of the components have yet to be devised, it is clear that a multibillion dollar investment package will be required. The estimated cost of the Kunming-Lashio Road System Improvement Project alone (R4), for example, is $817.4 million.

The countries of the GMS have decided on a phased approach to the development of the interregional transport system. A key guideline for the selection of the GMS road projects is that the projects will involve only existing transport infrastructure, that is, upgrading or "improving" existing roads and tracks. First tier, or priority projects, now under processing include three road projects: the Bangkok–Phnom Penh–Ho Chi Minh City–Vung Tau Road Project (R1), the East-West Transport Corridor Project (R2), and the Chiang Rai to Kunming Road Improvement via Lao PDR (Chiang Rai to Kunming Road) Project (R3). All three projects are at the feasibility level of project design. In May 1998, the Subregional Transport Forum provisionally agreed on a series of second tier projects. One of these involves road development, the Kunming to Hanoi Road Improvement Project (R5). In addition to interregional projects under the GMS Program, there are a number of medium and small-scale national level efforts to improve or expand existing transport networks. This is particularly true of the underdeveloped areas of Cambodia, Lao PDR, Myanmar, Vietnam, and Yunnan Province.

The decision to confine road development to upgrading of existing roads will help to limit social and environmental impacts. Nonetheless, the actual impacts will vary substantially, depending on the current condition of the road or track to be upgraded, the existing level of use as opposed to the projected level of use following upgrading, the remoteness of road sections from major population centers, and the ethnic and biophysical conditions adjacent to the road corridors. For example, the Bangkok–Phnom Penh–Ho Chi Minh City–Vung Tau Road Project (R1) will primarily facilitate existing trade links through areas of minimal environmental sensitivity. Conversely, sections of the East-West Transport Corridor Project (R2) and the Chiang

Rai to Kunming Road Improvement via Lao PDR (Chiang Rai to Kun-ming Road) Project (R3) have had minimal use in the past and will pass through some remote areas containing largely intact upland forests with high biodiversity conservation importance. Adjacent lands are populated by ethnic groups that have been largely isolated from mainstream development. As such, the potential for disruptive change is high.

Later in this chapter, we will address the Chiang Rai to Kunming Road Project as an example of an interregional road that offers great potential for furthering national and regional economic growth while also passing through areas of high biological and cultural diversity. The project will primarily involve upgrading of more than twelve hundred kilometers of road from northern Thailand through northern Lao PDR to Kunming, Yunnan Province. Of particular relevance to this chapter is the 250 kilometer section that passes from the Thai-Lao border through the remote northern part of Lao PDR to the Lao-PRC border. The existing road is not passable during the rainy season and therefore receives little use. Significant areas of intact forest and a population of diverse ethnic groups straddle this portion of the road. The Lao government has entered into a concession agreement with a private firm to improve this portion of the road for operation as a toll road. However, little progress has been made, and the ongoing feasibility study is addressing options for road improvements.

Socioenvironmental Considerations in Interregional Road Improvement

The Benefits of Road Improvement

Benefits expected from development of the three GMS first-tier roads are broadly similar and can be summarized as follows: improved economic and social interaction among the four countries; achievement of governments' development objectives and plans for agriculture, industry, and tourism in the influence area of the roads, thereby inducing accelerated industrial, tourism, and agricultural growth and furthering the distribution of social services to remote areas; reliable road links and less costly and time-consuming transport services; and poverty reduction through increased employment opportunities in areas along the roads. It is interesting to note that the general area of the Chiang Rai to Kunming road served as a subroute to the ancient Silk Road and in effect will promote trade objectives similar to those of two thousand years ago.

Experience from around the world confirms that expanding or improving a nation's road network can lead to a host of positive economic and social benefits. As one of the major tools of development,

Table 16.1 The Economic, Social, and Political Effects of Roads

Benefits (+)	Cost (−)
Facilitate Exchange Nation-nation Urban-urban Urban-rural Rural-rural	*Environmental* Landslides Erosion Siltation Hydrological Threats to wildlife and biodiversity
Reduce Transport Costs Increase efficiency Strengthen market forces Expand production	Uncontrolled pressure on finite land, water, and forest resources *Social* Cultural disruptions
Reduce Poverty Access to services goods Provide employment Access to market/credit	Follow-on migration Open access to lands Public health issues (e.g., pollution, disease) Exacerbation of inequalities
Environmental Natural resources management Provide access Relieve pressure on natural resources	Political tensions and conflicts

Source: Kirk Talbott, "Roads, People and Natural Resources: Toward a Regional Policy Framework for Transport Infrastructure in Montane Mainland Southeast Asia," Proceedings of the Montane Mainland Southeast Asia in Transition Symposium, Chiang Mai University, November, 1995.

the expansion or improvement of rural road networks has been one of the linchpins of economic development. Unlike many other "traditional" development activities, road building can provide substantial benefits to the rural poor (see table 16.1).

The combination of main trunk roads and rural feeder roads, for example, serves to increase agricultural productivity by facilitating access to inputs such as fertilizer, seed, and equipment. It also allows for surplus production to be moved to market, thus transforming traditional subsistence economies into more remunerative, cash-based market economies. A recent study in Africa, for example, highlighted the fact that the extension of feeder roads stimulates production by reducing the risk associated with the investment of scarce and precious capital, that is, time and inputs (Mwase 1995).

Improving rural road networks can better integrate otherwise marginalized communities into national, political, and economic structures by facilitating the exchange of goods, services, and information. By narrowing those gaps that retard the growth of isolated rural communities, the introduction of "modern" practices and knowledge can increase the well-being of millions of citizens.

The most obvious example of this is the increased availability of medical services, but that is just one of many improvements to the quality of rural life that is a direct result of the amelioration of rural transportation. One of the rationales behind the concerted donor effort to improve the road network in Cambodia in advance of the United Nations–sponsored elections in 1993, for example, was the more equitable distribution of national development benefits that would theoretically be forthcoming from an adequately represented rural electorate.

In addition, roads can further environmental conservation objectives. For example, they can increase access to government services that promote natural resource management techniques and inputs, such as community forestry extension, seedlings, and appropriate technologies. By increasing access to markets, capital, and credit resources, they can provide alternatives to unsustainable exploitation of natural resources.

The Costs of Road Improvement

There are, of course, liabilities to the construction and expansion of roads in developing countries. One of the weaknesses of planning for major road development from a social and environmental standpoint is the tendency to assume that these benefits will be a natural byproduct of road development. In fact, while increased economic activity inevitably occurs along roads (though not uniformly), in the absence of integrated planning it may disproportionately benefit a rather small group of economically astute people while placing additional social and economic strains on longtime residents, particularly indigenous peoples. Similarly, while roads may produce incremental benefits to the environment, when regional environmental protection is not factored into road planning, roads are likely to catalyze environmental degradation.

Decisionmakers responsible for environmental matters in the countries of the GMS recently identified transportation development as among the top four human-related activities that are expected to generate the most negative environmental changes during the next ten years (Mekong River Commission 1997). The initial environmental examination done for the Lao PDR portion of the Chiang Rai to Kunming road encapsulates many of the socioenvironmental issues associated with road improvements in remote upland sites in the GMS. (Table 16.2 summarizes the findings of the examination).

Generally speaking, these potential negative impacts fall into two broad categories: social and environmental (see table 17.1). In terms of the nature and timing of these impacts, both these broad categories can be further subdivided into three planning-level concerns: direct

Table 16.2 Potential Impacts

Direct	Indirect
Erosion and silt runoff	Loss of forest resources through illegal logging and "salvage" logging
Erosion from cut faces, shoulders, and borrow pits eroding into waterways	Current forest fragments are valued by residents and not currently under threat but could be opened to outsiders through road development
Erosion, vegetation loss, and impairment of downstream water quality from disposal of cut spoil	
Erosion of lands receiving concentrated outflow carried by drainage structures	
Impairment of fisheries/aquatic ecology	Impairment of fisheries/aquatic ecology
Changes to hydrology from erosion, landslides, bridgeworks, culverts, road runoff, and dumping of cut spoil	Loss or degradation of aquatic resources through increased pressure from licit and illicit fishing
Changes to river bed ecology from extraction of river sands and gravel for road construction	
Encroachment into environmentally sensitive areas	Encroachment into environmentally sensitive areas, e.g., protected areas
Loss of vegetation and habitat through road widening, realignment of the right-of-way, and extraction from quarries and borrow pits	Increased exploitation of resources by opening of new markets, increased ease of access to previously remote areas, increased population, etc.
Disruption or destruction of wildlife through interruption of migratory routes and other habitats	
Encroachment into built-up land and agricultural land	Loss of wildlife resources through increased pressure from illegal trade
Changes to irrigation structures from road alignment and sealing	The marketing of live and dead animals to Thailand, the PRC, and beyond
Removal of structures and crops encroaching on the road right-of-way	Capture of wildlife to supply local menageries
Destruction of agricultural land through road widening and realignment	Ease of wildlife hunting for local markets and food
Encroachment into cultural sites	Influences on cultural resources
Encroachment on known cultural heritage sites	Introduction of new people, lifestyles, goods, government influence, etc., may induce loss of cultural traditions
Encroachment on previously unidentified cultural heritage sites	"Opium tourism" has already started in some northern areas

Table 16.2 Potential Impacts (*continued*)

Direct	Indirect
Environmental aesthetics	Environmental aesthetics
Visual impact of road cut, spoil, disposal, and borrow pits	Increase in refuse, particularly non-biodegradable packaging
Environmental and social disruption by construction camps	Development of structures not compatible with surroundings
Dust nuisances	
Noise impacts	
Health and safety hazards	
Ground and water contamination by oil, grease, and fuel around gas stations and parking areas	
Toxic/hazardous materials spills from increased traffic	
Injuries and loss of life to people and animals from increased traffic volume and speed	
Creation of stagnant water in borrow pits, quarries, etc., suited to mosquito breeding and other disease vectors	
Air pollution	

impacts that occur during road construction and direct and indirect long-term impacts that begin to occur after construction with increasing road use.[1]

The most important of these three categories, and the one that can be most difficult to manage, is the long-term indirect impacts on environmentally sensitive areas such as forests and on previously isolated groups of indigenous people. Experience from major road projects in many developing countries around the world has shown that once these areas have been opened up, follow-on encroachment almost inevitably occurs. This may involve villagers searching for farmland or firewood or larger scale encroachment for commercial logging, mining, and other profitable enterprises. By opening up poorly accessible or economically underdeveloped areas, improved road infrastructure provides market access to both timber and nontimber products and a variety of agricultural goods that can be grown once the forest land has been cleared. Facilitated access also decreases both labor costs and the cost of transporting machinery and supplies, all of which may have contributed to making the proposed activity uneconomical before the construction of the road (Chomitz et al. 1995). Not surprisingly, past encroachment into environmentally sensitive areas has tended to be most intense in those areas where population pressures and low standards of living generate strong incentives to pioneer newly available land. Whether sponsored by the government or not, these activities have frequently exceeded expectations and put unsustainable pressures on the recently opened lands.

While the evidence remains scattered, development literature increasingly documents roads to be a major contributor to upland erosion.[2] The reduction of tree cover leads to increased levels of erosion, which in turn increases siltation, with negative repercussions on downstream water use for hydropower, agriculture, aquaculture, and consumption. Perhaps most serious, however, is that it diminishes policy and planning options for sustainable development when upland forests "ratchet down" biologically as they lose biomass, diversity, topsoil, and their complex structural and functional integrity.

In addition to the environmental costs are the concomitant social costs. The report of the Economic and Social Commission for Asia and the Pacific (ESCAP), *Environmental Impact Assessment Guidelines for Transport Development*, lists several quality of life values affected by highway/road development (UTESCAP 1990). These include involuntary resettlement, unequal distribution of access to markets, proximity to noise and pollution, and undermining of traditional social systems (such as tenure arrangements and culture).

Cultural disruptions can be particularly severe when minority or traditional communities are brought into rapid and prolonged exposure to "mainstream" culture. Given their generally marginalized status, these communities are frequently incapable of competing with the more savvy and politically astute people who come to profit from the new economic opportunities. Indigenous communities generally lose out in these competitions and may be further marginalized and impoverished, both economically and socially. And though roads facilitate the provision of much-needed medical supplies and expertise, they also have been the primary agent for carrying traditional diseases such as measles and influenza. They are now playing a more deadly role in bringing AIDS into previously uninfected areas in Asia.

Aesthetic, archaeological, and historical considerations often arise around road building and upgrading. Rural areas in Cambodia, Lao PDR, and Myanmar have a rich heritage of archeological, religious, and historical sites that are highly vulnerable to an uncontrolled infusion of outside people and interests.

Finally, the expansion of trade in this region of the world is not destined to include only government-sanctioned or licit activities. Whether intended or not, improvements in the transport sector in the GMS opens opportunities for development of a shadow as well as a legitimate economy, be it the smuggling of timber, gems, or other valuable commodities. Though these activities are extremely difficult to quantify or analyze, infrastructure advocates and planners must be aware of the entire range of activities that improvements in the transportation network will generate.

Integrated Planning

Improvement of national and interregional road networks that is done under the framework of an integrated regional development plan can improve prospects for achieving sustainable development objectives and can provide a range of benefits to local communities while protecting ecologically sensitive and biologically rich areas.[3] Experience in other regions of the world demonstrates the socioenvironmental costs of not doing so.

Examples abound of situations where poorly planned road projects led to unregulated or unplanned development that ended up degrading upland ecosystems and social structures.[4] Subsequent attempts to repair these "broken" upland ecosystems are generally an uncertain and extremely costly endeavor.

Although it involved the construction of new roads rather than improvement of existing roads, the case of the Transamazonia Highway is particularly instructive. For years, development agencies pushed for large-scale road building projects in western Amazonia as a way to bring lifestyle-improving economic activities into previously inaccessible areas. Once these roads had been built, however, government agencies found that they had little control over the volume and nature of migration. Employment seekers swarmed into newly accessible areas much faster than had been expected, quickly outstripping the provision of services and amenities.

As a result, the unique and valuable forest system was severely compromised by the follow-on activities often associated with road construction: logging, ranching, and small-holder agriculture. The cycle was repeated as soil nutrients became exhausted and even more lands were opened up by feeder road networks. As a similar scenario begins to unfold in Central Africa, the same adverse consequences are being experienced in that large, heavily forested region.[5]

Environmental Impact Assessments (EIAs) and Social Impact Assessments (SIAs)

Project level EIA and strategic environmental assessment (SEA) are perhaps the most important tools available for predicting and addressing the potential impacts of road development policies, plans, and projects.[6] They can identify anticipated environmental and other costly impacts of ecologically fragile landscapes before the project gets underway. That allows decisionmakers to balance the demands of immediate gain from economic investments with the long-term environmental and social consequences of the project. If initiated early, carried out properly, and integrated conscientiously, they can preclude the need for expensive remedial actions later on.

The EIA experience in countries of the GMS varies widely, while SEA remains a virtually untried tool. Thailand and the PRC have developed their EIA systems over approximately two decades and now have reasonably strong systems with cadres of trained professionals. Conversely, Cambodia has only recently instituted an EIA system, and Lao PDR has yet to formally institute comprehensive EIA guidelines and procedures at the national level. While EIA theory advocates interjecting environmental considerations into the earliest stages of project development—when the project design can be changed to avoid or lessen environmental impacts—in practice many EIA studies fall short of this goal. Rather, EIA studies often are conducted after key project decisions have been made. Even in countries with a long tradition of environmental assessment, the project level EIA often does not reach far enough upstream in the development planning process to strike a clear balance between economics and environmental protection.

Given the high population densities in much of the GMS and the fact that many of the inhabitants of areas targeted for road development are ethnic minorities and other vulnerable groups, SIAs are another critical component of transport sector development planning. By anticipating the consequences of resettlement and dislocation that can be caused by road building, SIAs can not only help planners mitigate the adverse social impacts associated with road expansion, they can also allow them to maximize the social benefits.

All road projects currently being designed under the GMS Program incorporate the ADB's *Handbook for Incorporation of Social Dimensions in Projects* (ADB 1994). Specifically, the handbook calls for the identification of project-affected people, especially those deemed to be most vulnerable; an assessment of their needs and demands (including gender and class considerations); and an appraisal of the absorptive capacity of the environment. ADB projects require an initial social assessment, preferably during the early design stage, and a full social assessment if the initial assessment indicates the need.

EIA and SIA work best when there is meaningful public participation during project planning and design. Here again, the means, methods, and capabilities for designing and implementing public participation programs vary widely among the countries of the GMS. As a general observation, the record for public participation in road projects has been weak.

Institutional Capacity

The process of designing, constructing, and maintaining socioenvironmentally sound road projects requires political and community-based support, institutional backing, accurate data, the delineation of clear roles, adequate quantities of skilled human resources, and effective enforcement policies. Unfortunately, the national agencies charged with environmental regulation in most of the countries of the region are of quite recent establishment. As a result, they tend to be both weak and inexperienced. The more compromising effects of this current status include:

- Poor utilization of research results, inventories, Geographical Information System (GIS) products, and environmental data in environmental natural resource management planning and policy formulation; the intended users of data are often not identified and consulted during research planning, so the connection between research and development planning is weak. Vietnam, for example, has a wealth of information that should enhance the policy process by clarifying the range of options available. But this typically does not happen because the information often comes in indigestible forms, is not timely, and/or does not have sufficient follow-through support.
- Few sound empirical indicators that allow for the monitoring of trends as economic growth and infrastructure development in the region accelerates; there frequently is a lack of confidence in existing indicators that measure social and environmental changes; hence, few systemic approaches use indicators effectively in monitoring the activities related to road building.
- Low capacity for strong partnerships between government agencies and a variety of private or informal entities to compensate for inadequate and/or diminishing human resources in the public sector; unfortunately, there are few local nongovernmental organizations (NGOs) and independent "think tanks" to assist government in formulating policy related to transport and other large-scale infrastructure development. Thailand is the only country in the region that has a significant independent sector. Its impact on government policy has been significant, and NGOs have proven to be an essential part of the policymaking process in Thailand.
- Generally insufficient local level participation in the management of vulnerable natural resources where roads and other infrastructure projects are being built; one of the most promising strategies to stabilize upland forests resources is through the creation of partnerships between rural communities and government forest agencies. But since the policy, legal, and implementation frameworks to do this are generally not well developed, little action has been taken to date in several countries. Of special concern in regard to upland forests are tenure issues and the need to decentralize resource management systems.

- Poor coordination among line agencies and overlapping mandates; conflicting initiatives create field-level implementation programs that squander resources and undermine effective multistakeholder management practices.
- Insufficient "horizontal" communication within the region. Too much emphasis is placed on generating information that is tailored to the needs of funding agencies or research institutions based outside the GMS. While some excellent networking activities are underway, linkages between countries remain weak.

Strengthening Prospects for Socioenvironmentally Acceptable Roads

Roads are the most invasive, and important, infrastructure development in rural areas of the GMS, where conflicts with environmental protection and social stability will be most challenging. A number of activities are ongoing or planned that will help to address the constraints to environmentally sound road development in the GMS.

Integrated Regional Planning

Potentially the most important initiative taken thus far under the GMS Program to promote closer linkage between economic development and socioenvironmental considerations is the recently approved technical assistance for the "Strategic Environmental Framework for the GMS." The technical assistance will assist the six GMS countries to prepare a strategic environmental assessment for the subregion that will promote the upstream integration of environmental and social considerations in economic development planning. The assistance will aid in this assessment by identifying and characterizing areas of potentially serious conflicts between socioenvironmental and development interests, such as the development of road systems that traverse protected areas, or areas that require special development planning and management approaches because of sensitive environmental and social conditions. Options for avoiding these conflicts or introducing special planning and management approaches will be recommended. Although the entire range of development sectors will be taken into account, special attention will be given to road and hydropower development. Recommendations for strengthening participatory planning techniques in GMS development will be among the main outputs of the technical assistance.

Another regional planning initiative about to get underway is the Poverty Reduction and Environmental Management in Remote GMS Watersheds program. This initiative aims to (1) analyze the various

GMS country strategies and policies related to watershed management and, based on the analysis, to make recommendations that will harmonize them across the subregion; (2) prepare a subregional policy framework for sustainable watershed management, with special emphasis on the nexus of rural poverty and watershed degradation; and (3) prepare feasibility studies for watershed investment projects at selected priority sites. This program potentially will play an important role in identifying key upland forest areas in the vicinity of regional roads for protection and management.

A third regional initiative that will influence road development is the GMS Working Group on Environment. Made up of senior environmental and natural resource policymakers from each GMS country, the Working Group has targeted the proponents of road development for special dialogue. Consequently, Working Group members are about to become more involved in the early planning stages of interregional roads. As an initial effort, a paper to identify critical issues in controlling cross-border traffic of endangered flora and fauna is now under preparation. Working Group members also have participated in meetings of the GMS Transport Forum to strengthen the attention given to environmental and social issues.

Environmental Impact Assessment

Efforts to provide strategic socioenvironmental planning direction to road development at the regional level will take time to yield results. In the meantime, continuing emphasis on strengthening EIAs is essential. A number of aid agencies have provided assistance to strengthen national EIA capacity in GMS countries. The ADB is currently formulating assistance to Lao PDR to develop a national EIA system and to strengthen the environmental management capability of major line agencies, including the Department of Communications, which is responsible for road construction. This is essential for this strategically located country, through which several of the regional roads will pass. EIA training programs are continuing in Cambodia and the PRC. Environmental and social assessments are being done as integral steps in planning for the interregional roads now under prefeasibility or feasibility design.

The Chiang Rai to Kunming Road Project provides an indication of how EIA and SIA can help to change the planning of road development. The initial findings of the ongoing feasibility-level investigations have highlighted the need to expand the scope of the project beyond purely engineering concerns and impact mitigation. A broader development view is being recommended that would encompass opportunities for socioeconomic development and environmental protection in an area that extends out from the road corridor to outlying lands and communities.

From the socioeconomic perspective, the recommended design recognizes the need to support programs for community preparation so that residents are able to anticipate and adjust to the lifestyle changes that will inevitably occur. But it also takes note of the need to clearly factor in the presence of diverse cultures, including the comparative advantages and disadvantages that these people will have once road improvement is completed. Consideration is being given to the provision of social and cultural amenities that will be required by these communities, such as schools, health care, clean water supplies, cultural conservation programs, handicraft promotion, and others. Consideration also is being given to how new income opportunities can be planned rather than just allowed to happen.

The feasibility study is considering a number of initiatives that can lead to improved environmental protection and management. Among these are support to (1) develop local sustainable marketing of non-timber forest products, providing a potential incentive for sustainable management of forests; (2) establish and provide management support for new protected areas; (3) initiate programs for livestock husbandry improvements that will decrease pressure on forests; (4) expand fish culture to draw pressure away from fish capture in local rivers; (5) foster a community-based forest management and control regime that will inhibit outsiders from exploiting forest resources; and (6) encourage local involvement in sustainable commercial forestry practices.

Institutional Capacity

In addition to EIA strengthening, a number of activities are ongoing to increase institutional capacity as part of the GMS Program. Over $11 million has been directed at environmental technical assistance under the GMS Program to date. In the field of information exchange, support has been provided to develop a Subregional Environmental Monitoring and Information System (SEMIS). The ultimate goal of SEMIS is to base GMS resource allocation and environmental management decisions on current and accurate information. The project is establishing procedures for the six GMS countries to share environmental information and to establish a subregional environmental database that will be accessible to researchers and decisionmakers alike by using standard data and metadata formats. SEMIS results have been so encouraging that the Association of South East Asian Nations is considering adoption of the SEMIS framework.

The Subregional Environmental Training and Institutional Strengthening (SETIS) project aims to assist the GMS countries to formulate and implement environmental policies, legislation, and programs of common significance; provide a forum for the governments to share their experiences; provide training in areas of critical environmental concern from a regional perspective; and formulate mutually agreed-upon environmental standards and programs. Training sessions are being provided for a range of topics, such as EIA, risk and hazard assessment, strategic environmental assessment, cumulative impact assessment, integrated resources management, and others. An important outcome of SETIS has been the formation of peer group linkages among environmental professionals in the six countries.

Nationally, the ADB is assisting Lao PDR in developing a system that will introduce techniques for bioregional (watershed/river basin) planning to improve the integration of development planning and implementation with social and environmental concerns. This is a new concept for the government, but the early enthusiasm shown for such planning by the participating provincial governors provides a glimmer of optimism that this concept may gain a foothold in the country.

The ADB and governments are increasingly encouraging NGOs and other nonformal sector agencies to become involved in the GMS program. NGOs are regularly invited to attend Working Group meetings, and more emphasis is being given to involving the public in the design, implementation, and monitoring of technical assistance and investment projects. A recently concluded technical assistance study canvassed the NGO communities in three of the six GMS countries (Cambodia, Lao PDR, Vietnam) to look at how the ADB can further cooperative activities with NGOs.

Areas for Improvement

A good start has been made at better integrating socioenvironmental concerns into road development programs. The strategic environmental framework study, in particular, is a potentially major step toward moving socioenvironmental concerns nearer to the early planning stages of infrastructure development before firm decisions are made on site selection and other important parameters. It will be as important for the ADB to take on board the lessons learned from the study when it develops its country assistance programs as it will for the GMS countries themselves.

However, there would appear to be two major areas where the GMS Program is still particularly weak. The first is public participation. While efforts are being made at the project level to increase opportunities for project stakeholders to learn about projects and to meaningfully participate in their development, the efforts are somewhat spotty and are not consistent throughout the sectors. Methods need to be devised and implemented that will make information available in a timely and understandable way. This is a formidable task, as it requires working with several different languages and several different national approaches to public participation. The benefits of a well-informed and active body of stakeholders would appear to be worth the investment in time and money that an effective public participation program will require.

A particularly valuable segment of stakeholders is the NGO community, both international and local. NGOs are frequently close to rural populations that are affected by road projects. This offers the advantage of using NGOs to spread information over a wide rural area and to assist rural stakeholders in articulating their views to the road proponents.

Technical support to strengthen NGOs as both effective advocates for local and national interests and as participants in project development and monitoring can pay dividends by strengthening public support for socioenvironmentally friendly road projects and by ensuring that the long-term impacts of the roads are minimized. Gone are the days when the anticipated economic benefits of large-scale projects were left to sort themselves out according to local-level dynamics. NGOs can help the ADB and other aid agencies to maintain their accountability for projects during and after construction by serving as independent monitors.

Additionally, there is the often overlooked role of the private sector, especially regarding follow-on economic activities. In appreciation—and anticipation—of this role, national and regional policymakers need to consider new regulatory mechanisms and incentive structures that will minimize adverse environmental and social effects of private sector activities. Unless such considerations are integrated into a comprehensive planning and implementation framework, much of the increasingly good work done by public and aid organizations may be undermined.

A second major problem area is quantifying all of the costs and benefits associated with road projects. It is important to quantify the positive and negative, direct and indirect, short-term and long-term socioenvironmental impacts of road building (and other infrastructure development), particularly in rural areas where communities continue to rely heavily on natural products for their sustenance.

At the local level, more consideration can be given to environmentally friendly road improvement methods. Interregional roads need not be synonymous with large highways. The road grade can have a significant bearing on how much direct impact the road will have on forest resources, cultural sites, and communities. Feeder roads can be carefully planned to minimize impacts. In Nepal, the ADB is supporting the "low cost, environmentally friendly, self-help, or LES," approach to road construction that relies on a minimum right-of-way, balanced cut-and-fill, minimal use of explosives and heavy equipment, strong local beneficiary input to construction and maintenance, and technical training.

A continuing commitment to capacity building is another important aspect. In this new era of small government, the provision of technical assistance to government agencies must be highly selective and strategic. Comparatively more emphasis might be given to developing partnerships between government agencies and organizations with special skills required by these agencies. Of particular relevance in this regard is the wealth of regional and extraregional experience that can be tapped to introduce a variety of cost-effective, cross-pollination programs. Examples abound; just a few will be mentioned here.

More opportunities might be taken to support study tours to project sites in Asia (Thailand, India, Nepal, and Pakistan, for example) where practical lessons can be learned quickly and dramatically. Formal linkages could be made with leading academic centers in the region such as Chiang Mai University (CMU), the Yunnan Institute of Geography, the Asian Institute of Technology, and the Center for Natural Resources and Environmental Studies at the University of Vietnam in Hanoi.

By collaborating with leading international policy research institutions such as the East-West Center in Hawaii and the International Centre for Integrated Mountain Development (ICIMOD) in Kathmandu, helpful Asia-specific insights can be added to the knowledge base. The East-West Center has been working with government and NGOs throughout the region for many years on a wide variety of critical environmental and developmental issues.[7] CMU has been a leader in addressing the challenges of upland development and resource management. In November 1995 CMU hosted a symposium, "Montane Mainland Southeast Asia in Transition," which was a highly successful event in galvanizing attention among researchers and development practitioners in the region.[8] ICIMOD has more than a decade of experience in addressing development, environment, and social issues in severely sloping lands, including road development.

GIS can be an important tool in road planning, particularly when roads are designed as components in an overall regional development plan. As mentioned earlier, GIS assistance is being provided under the GMS Program. In addition, numerous multilateral, regional, and bilateral organizations are providing GIS assistance to various government agencies within the GMS. Perhaps what is most urgently required is improved coordination in this tangled web of assistance, and more emphasis on building local capacity to effectively use GIS analytical tools.

A recent World Bank study showed how GIS can be used to quantify the anticipated environmental and economic impacts of rural roads in Belize, another tropical country with significant upland regions (Chomitz 1995). GIS databases covering land use/land cover, soil suitability, land tenure, and distance to roads and market were assembled.[9] The model showed that market access and distance to road strongly affect the probability of cultivation. This allowed planners to assess the relative costs and benefits of greater improvement and use of existing roads as opposed to building a sprawling road network.

Analysis of the relationships between land use and distance to road/on-road travel time showed that building a feeder road in agriculturally suitable land to points that were two kilometers from the main road more than doubled the probability that the land will be cultivated. Conversely, building a feeder road into marginal land had relatively little impact on the probability of agricultural use. The study suggests that there are high economic returns for increasing the density of roads in productive areas near market centers. In contrast, building roads in more remote areas may yield few economic gains but have substantial environmental consequences.

Another area that could benefit from more attention is the role that tenure policy plays in affecting the actions of upland forest dwellers. The ongoing feasibility study of the Chiang Rai to Kunming road, as explained earlier, is considering active participation from local communities in social development and resource conservation programs as part of the road improvement project. Without secure and clear land tenure arrangements, however, these good intentions are unlikely to produce the desired results.[10]

Conclusion

Not surprisingly, given their shared biogeophysical context and their similar historical antecedents, the countries of the GMS face many common opportunities and constraints in dealing with their socioeconomic development aspirations and environmental protection imperatives. The six countries also share a common long-term objective: the sustainable use of the natural resource base so that it may continue to provide critical environmental services, valuable natural products, and vital living space for millions of the region's inhabitants.

A concerted effort is necessary to achieve these objectives so that the countries prosper together, while avoiding economic and environmental externalities that may burden their neighbors within the GMS. Nowhere is this concerted, regional effort more important than in the planning and execution of infrastructure projects such as roads. Significant and deliberate strides have been made in this direction within the framework of the GMS. Much more remains to be done.

An Action Plan covering the years 1998–2000 that is now under preparation recognizes many of the lingering shortcomings of the GMS Program. It calls for improvements in project evaluation and monitoring, and it recommends enhancements in information dissemination activities. It also suggests the need for revisiting the premises and assessing the changing environment of the Program. This chapter has offered some thoughts on how the countries and the ADB might better integrate socioenvironmental considerations into the GMS road development program.

NOTES

1. For a more detailed analysis of these potential adverse impacts and effective mitigation activities, see *Environmental Assessment Sourcebook*. Vol. 2. Sectoral Guidelines, World Bank Technical Paper Number 140. The World Bank, Washington D.C., 1991.
2. A recent field study in Thailand found strong evidence that unpaved road surfaces may be the primary source of runoff and sediment in recently deforested catchments. Cited in Giambelluca et al. (forthcoming).
3. For a detailed, theoretical description of the factors at play here, see Bayliss (1992).
4. For a detailed look at the dynamics of this interplay, see Deoja, (1994).
5. For a detailed discussion of the human dynamics of land use change in Central Africa, see WWF, Nature Conservancy, and WRI (1993: chap. 3).
6. For an overview of the evolving role of EIAs, see Terivel et al. (1992). For a more complete discussion of the role that EIAs can and do play in Asia, see Smith and van der Wansem (1995).
7. For a detailed look into the East-West Center's environmental and development issues in the region, see Rambo et al. (1995).
8. See "Montane Mainland Southeast Asia in Transition," Chiang Mai University, February 1996. This chapter builds on an earlier paper submitted by Kirk Talbott at this symposium.
9. In addition, a multinormal logic regression was run to measure the association between these variables and three land-use types: natural vegetation, semisubsistence farming, and commercial farming. Sampling was carried out using a three-kilometer grid.
10. For a more complete discussion of the role that traditional forest-dwelling communities play in the protection of forest resources in seven Asian and Pacific nations, see Lynch and Talbott (1995).

REFERENCES

Brunner, Jake, Kirk Talbott, and Chantal Elkin. 1998. *Logging Burma's Frontier Forests: Resources and the Regime*. World Resources Institute, Washington, D.C.
Bayliss, Brian. 1992. *Transport Policy and Planning: An Integrated Analytical Approach, EDI Technical Materials*. World Bank, Washington, D.C.

Chomitz, K., et al. 1995. *Roads, Lands, Markets, and Deforestation: A Special Model of Land Use in Belize*. Policy Research Working Paper no. 1444. Policy Research Department, Environment, Infrastructure and Agriculture Division, World Bank, Washington, D.C.

Deoja, Brenda B. *Sustainable Approaches to the Construction of Roads and other Infrastructure in the Hind Kush-Himalayas*. ICIMOD Occasional Paper no. 24. International Centre for Integrated Mountain Development, Kathmandu.

Economic and Social Commission for Asia and the Pacific. 1990. *Environmental Impact Assessment: Guidelines for Transport Development*. ESCAP Environment and Development Series. United Nations, New York. p. 13.

Giambelluca. Forthcoming. *Hydrology and Landscape Evolution in Mountainous Tropical Watersheds: The Impact of Rural Roads*.

"Handbook for Incorporation of Social Dimensions in Projects" Asian Development 1994.

Lynch, Owen J., and Kirk Talbott. 1995. *Balancing Acts: Community-Based Forest Management and National Law in Asia and the Pacific*. World Resources Institute, Washington, D.C.

MacKinnon, John, et al. 1986. *Review of the Protected Area System in the Indo-Malayan Realm*. International Union for the Conservation of Nature and Natural Resources, Gland, Switzerland.

Mekong River Commission. 1997. *Greater Mekong Subregion State-of-the-Environment Report*. Chiang Mai University Consortium. 1996 *Transition*, 12-16 November 1995, Chiang Mai, Thailand. Bangkok. June.

"Montane Mainland Southeast Asia in Transition," Chiang Mai University, February 1996.

Mwase, N. 1991. Role of Transport in Rural Development in Africa. Impact of Science on Society, 41:2 (1991), pp. 137–148 as quoted in Francis Dillon. "GIS and Transportation Planning in Sub-Saharan Africa: Accommodating Village Centered Transport Tasks" (Draft), George Mason University, Fairfax, Virginia, October 1995.

Rambo, A. Terry, Robert R. Leed, Le Trong Cuo, and Michael R. Di-Gregorio. 1995. *The Challenges of Highland Development in Vietnam*. East-West Centre, Hanoi University, University of California at Berkeley, October.

Smith, David B., and Mieke van der Wansem. 1995. *Strengthening EIA Capacity in Asia: Environmental Impact Assessment in the Philippines, Indonesia and Sri Lanka*. World Resources Institute, Washington, D.C.

Social Dimensions Unit of the Asian Development Bank, 1994. *Handbook for Incorporation of Social Dimensions in Projects*. Asian Development Bank, Manila.

Talbot, K. 1998. Logging in Cambodia: Plunder for Power. *Cambodia and the International Community: the Road Ahead*. (The Asia Society and SAS). June.

Terivel, Riki, Elizabeth Wilson, Stewart Thompson, Donna Heaney, and David Pritchard. 1992. *Strategic Environmental Assessment*. Earthscan Publications, London.

World Bank. *Lao PDR: Environmental Review*. East Asia and Pacific Regional Office, Agriculture and Natural Resources Operations Division 1993.

WWF, Nature Conservancy, and WRI 1993. *Human Interaction with the Forests of Central Africa in Central Africa: Global Climate Change and Development.* Technical report. Biodiversity Support Program, World Wildlife Fund, the Nature Conservancy, and World Resources Institute, Washington, D.C.

17 Roads and Tropical Forests

From White Lines to White Elephants

John Reid

When conservationists debate the underlying causes of habitat loss in the tropics, roads often top the list. Roads open formerly inaccessible wilderness areas to any human activity that can be practically undertaken. It is precisely for that reason that roads are so popular and present such a formidable conservation challenge. This chapter examines the nature of roads' environmental impacts and presents four fundamental strategies to ensure that road planning minimizes the worst and least reversible of these impacts. Recent experiences in Brazil and Bolivia illustrate how the threat of road-induced habitat loss was addressed in varying circumstances.

The Problem with Roads

If there is one road project that awakened concern over the connection between roads and deforestation, it was Brazil's BR-364. Financed by the World Bank, the Inter-American Development Bank (IDB), and the Brazilian government, this highway links the Amazonian states of Mato Grosso,Rondonia, and Acre. The road cut through wilderness, biological reserves, and indigenous territories and ignited bloody conflicts between ranchers and rubber tappers; spectacular photographs and satellite images of the concomitant forest destruction were beamed around the globe.

P. M. Fearnside (1987) reported that deforestation in Rondonia climbed from a background level of 1,216 square kilometers in 1976 to 13,955 square kilometers in 1984, the year the pavement reached Porto Velho, Rondonia's state capital. Fearnside's study, though a valuable chronicle, didn't distinguish between the road and other causes of habitat loss. A series of other factors—notably, large official subsidies for agricultural colonization—were also motivating deforestation at the time. A. Pfaff (1997), however, showed that roads, especially paved ones, have been central factors in Amazonian deforestation episodes. In a statistical analysis of satellite data from 1978 to 1988, he found that among variables over which policymakers exercise some control, road paving was the one most strongly associated with deforestation. Construction of unpaved roads had a similar, though less pronounced effect. Other studies have shown similar associations between road infrastructure and forest loss in Belize (Chomitz and Gray 1996), Guatemala (Sader et al. 1994; Sader 1995), Thailand (Cropper et al. 1997), and Cameroon (Mamingi et al. 1996).

Roads provoke habitat loss because, more than any other type of transportation infrastructure, they form the loci of human settlement. There are several characteristics that place roads in this role, as follows.

Multiple purposes. Unlike railroads, airports, or waterways, roads accommodate the widest possible range of human locomotion, from the largest trucks and buses, down to animal and pedestrian traffic. This democratic characteristic allows people of various means and with various economic purposes to penetrate formerly inaccessible places.

Low variable costs. The fixed costs of road transportation are large. Building them, particularly over rough terrain, is extremely expensive. Air transportation, by contrast, has low fixed costs—the pavement needed may equal no more than several kilometers of road. Roads' variable costs (largely the cost of operating vehicles), however, are very low—another factor that makes road transport available to a wide array of social and economic groups.

Expandability. Road systems are almost endlessly expandable. Roads of progressively smaller size and lower quality tend to be appended to any trunk road and may reach far into the hinterlands. Road size and quality can be scaled to the purposes or communities they serve and the vehicles needed to reach them. Such expandability is constrained in the case of air and rail travel, by minimum limits on the quality of the infrastructure, and in the case of water travel by the more or less given natural infrastructure of waterways.

In the face of all these advantages for extending and serving human settlements, people have tended to invest in cars, trucks, and buses rather than boats, trains, and airplanes, further reinforcing the demand for new roads.

How to Do Roads Right: Four Rules of the Road

Despite their potential to alter natural landscapes, or because of it, roads will continue to be built and improved on a grand scale. That reality raises questions: Is there a way to do roads right? Are there ways of analyzing transportation goals and conservation priorities and planning infrastructure projects so that they do less ecological harm, or actually do some good? In this section, I outline four general "rules of the road" planners can use to improve conservation outcomes related to roads.

Avoid sites within high biodiversity and vulnerable intact ecosystems.

Any improvement in access to a region will unleash a chain of consequences that, after a certain point, are impossible to control or predict. Therefore, in harmonizing transportation needs with habitat conservation goals, the first line of action is to simply try to keep the roads far away from sensitive ecosystems. To do so, regional and national planners must take stock of biological resources early in any planning process, before certain projects or approaches gain political momentum and take on a life of their own. An example of such an effort is a collaboration between the Andean Development Corporation (CAF) and Conservation International (CI), in which conservation priorities were cross-referenced on maps with transportation priorities.

Don't build roads whose net economic benefits are likely to be negative.

The economic merit of a project may seem unrelated to habitat impacts, but it's not. That's because projects in sparsely populated hinterlands have fewer beneficiaries and less potential to unleash economic activity than do projects in more densely settled areas. In a study of potential road projects in the Bolivian Amazon, none of the planned routes convincingly passed a cost-benefit test using the development banks' standard economic model. The most remote of the three was the worst performer, with a projected internal rate of return (IRR) of 7 percent (in Bolivia, development banks generally require an IRR of positive 12 percent to judge a project economically beneficial) (Reid and Landivar 1997).

Officially, any project funded by a major development bank will have to have made a case that its transportation benefits will exceed its costs. Unfortunately, state-funded projects are often free of such constraints, and even bank-funded projects may be approved on the basis of unreasonably optimistic assumptions, especially regarding the costs of construction and maintenance and the projected numbers of users. Independent assessments of economic viability can therefore be an essential tool for conservationists (see the Madidi case study described hereafter).

Scale quality to need.

Any major road development that does take place should be scaled to the purposes for which it's being undertaken. For example, paving roads to provide access for sustainable forestry is probably a counter-productive strategy. On the one hand, paving reduces the cost of log transport and would make it economically feasible to extract tree species of lower value (for a discussion, see Rice et al. [1997] and Reid and Rice [1997]). However, the improvement makes other land uses more competitive relative to forestry. The lack of good road infrastructure confers a comparative advantage on forestry for several reasons. First, timber is relatively nonperishable, so transport time and reliability are less important than they are in the marketing of agricultural goods. Second, loggers don't need to establish permanent settlements in the forest, so they can do without the nonroad infrastructure and services often made available by paved roads. Finally, the nature of timber extraction requires that loggers possess equipment to build roads, whether or not a paved trunk road exists. In remote areas, loggers tend to keep roads and bridges passable as needed, using their own equipment, thereby exercising some control over access.

Focus environmental planning on the most serious and irreversible impacts.

Roads have several different kinds of environmental impacts. The most direct and obvious is that they chew through whatever natural vegetation and features may be in their path. The actual area disturbed, however, is slight, and the bisection of habitat can be overcome by most species. A more serious direct impact is the erosion of road cuts and even road surfaces, and the resulting deposition of excessive sediment in streams. Smaller streams also suffer from the failure to use adequate culverts and other devices to allow them to pass under roads. The most serious and irreversible impacts are indirect: large-scale changes in land use that have nothing to do with the construction of the road but a lot to do with the people who travel on it.

While it's clear that deforestation induced by roads is far and away the largest impact in many places, it has traditionally received the least attention. For one thing, planners may not consider it an undesirable impact since they may favor colonization. But even where deforestation is recognized as something to be avoided, the institutions building the roads almost never have expertise, inclination, or incentives to do anything about it. Public works agencies are staffed by engineers, not biologists, and are judged on the efficiency and quality of their construction efforts, not their ability to avoid and preserve critical habitat. Even where environmental specialists are placed in such agencies, the institutional culture and goals do not confer very much influence on them.

The first step to controlling any potential habitat loss is to establish or strengthen strategically placed protected areas that ensure ecosystem protection once the area becomes more accessible. This strategy has proven useful in Costa Rica, where the Braulio Carrillo National Park protects rain forest bisected by the road linking the central valley to the Caribbean coast. It also has worked in Venezuela and Paraguay and is now being used in a section of Brazil's highly endangered Atlantic coastal forest (see the Itacare case study hereafter).

The specific protected areas strategy employed should match the circumstances of each road. In a region with both well-protected and unprotected forest, it may be advisable to run the road through the protected area, where land use regulations are strict and simple. In other places, a declared park may need to be equipped, staffed, and consolidated before it can serve the purpose of mitigation. In other circumstances, a new park may need to be created to fulfill the protection function.

For protected areas to succeed in deflecting indirect road impacts, other government policies and plans must be harmonized with the road's environmental planning. Amborû National Park, for example, was established in part to mitigate the impact of planned road improvements between the Bolivian cities of Cochabamba and Santa Cruz. The road, however, was the focus of government-sponsored colonization efforts that brought waves of migrants to the vicinity of the park. As a result, many of the more accessible, low-elevation areas of the park have been turned over to farmers.

The Case of Itacare: Build the Road, Protect the Forest

A road that exemplifies a successful marriage of conservation and transportation goals is the coastal blacktop linking the Brazilian beach towns of Ilheus and Itacare. This region was once part of a vast belt of rain forest that stretched the length of Brazil's Atlantic shore. Today this corner of the state of Bahia is one of the very few places where the lowland coastal forest remains. A profusion of endemic species survives in the local forests, including three unique primates.

In 1995 a team of conservationists from CI and the Institute for Social and Environmental Studies in Southern Bahia (IESB) learned of plans to pave a road into one of region's least accessible towns, Itacare. The team was conducting research somewhat to the south and, beyond anecdotal accounts, was unsure of the extent of forest cover remaining in the vicinity of the proposed road. A year earlier, a local sawmill owner had mentioned to the team that a large tract of forest remained on a series of ridges known as the Serra do Conduru. One forest studied on the flanks of those ridges by a different group of scientists in 1993 had yielded the highest per-hectare diversity of plant species on Earth, with a dozen species new to science (Thomas et al. 1998). After reviewing satellite images and driving the dirt tracks in the area, the team confirmed that one of the last major conservation opportunities in the Brazilian Atlantic forest was at hand.

A complex set of political and institutional factors faced the CI/IESB team. Municipal elections were coming up, and the road enjoyed rapturous local support. Furthermore, the segment of pavement was part of a much larger statewide ambition to make the whole coastline accessible to tourists. This plan was, in turn, a piece of a grand and popular scheme, backed by the Brazilian government and the IDB, to stimulate the tourism industry in the country's northeast, a poor, sunny region well-endowed with beaches. Only one local environmental group, called the Black Dolphin, had openly criticized the road plan. Politics aside, there was a fairly solid economic rationale for the road. Pavement was needed to draw tourists in passenger cars into a region whose stunning beauty and beaches made tourism a logical economic development choice.

At first the CI/IESB team merely alerted state officials and the IDB to the biological resources at risk in the region, and received assurances that a proper environmental impact assessment (EIA) was being prepared. Next the team consulted IDB documents and experts to see, in the likely event the road could not be stopped, what might constitute a credible plan to avoid and mitigate its environmental impacts. Those inquiries indicated the need to counter the deforestation potential of the road directly, by protecting what was left of the forest in an official reserve.

The finished EIA made no reference to the threat to habitat, focusing instead on the typical set of direct impacts of moving around large quantities of dirt and rock. The nongovernmental organization (NGO) team began the task of selling the idea of a road and a park to the state, the IDB, local mayors, and environmentalists like those at the Black Dolphin. IDB's environmental officer visited the area, criticized the quality of the EIA, and seconded the park as a condition of approving the road funding. Within the state government a newly formed and active forestry agency, which would oversee any park created, stepped in and also backed the idea. Local environmental activists also endorsed the plan, after coming to understand the inevitability of the road and to recognize the small window of opportunity to acquire and protect the surrounding forest.

That combination of supporters was sufficient to overcome reluctance in other quarters of IDB and impatience within the state road-building agency. While the IDB proceeded to shuffle funds to enable purchase of the land, the state forestry director set to work to win the governor's approval, and the CI/IESB team carried out mapping and technical design tasks for the new park. Early in 1997 the governor of Bahia declared the Serra do Conduru State Park.

Several lessons were underscored by the Itacare experience. First was the importance of focusing on the largest, most irreversible impact of the road—deforestation. Second was the importance of having planning responsibility vested outside the public works agency. While the road builders weren't interested in the park, they also did not oppose it. It just wasn't any of their business. Fortunately, there was a separate planning agency—the state tourism bureau—in charge of implementing the IDB loan. The planning agency provided neutral ground on which to integrate the conservation and road plans. Another lesson was the value of a three-way partnership that brought together the authority of government, the financial resources of the lender, and the local political and biological knowledge of the environmental NGOs.

The Case of Madidi: White Elephant Habitat

The Madidi region is one of the most remote places in Bolivia. It is true wilderness, straddling the Andes foothills and the Amazon basin in the northwest part of the country. Due to poor access, it has remained sparsely populated and boasts superlative biological resources (CI 1991). The biological importance of the region was recognized officially in 1995, when the Bolivian government created the Madidi National Park, which embraces 1.8 million hectares of lowland and upland forest and savanna. Road development here presents conservation challenges that are common to frontier areas throughout the tropics. The approach taken here to those challenges was to analyze the economic merit and biological risks of road development in the area.

Currently the only road of note in the area is a gravel route skirting the foothills and roughly running along the border of the national park, northwest toward the Peruvian border. The quality of the road declines as it presses toward Peru, and it is impassable past the Madidi river, where there is no bridge. It runs right through the beds of the many streams coming off the mountain range on its way to several small towns.

In the early 1990s, the Bolivian government submitted a proposal to the Corporación Andina de Fomento (CAF), a regional development bank, to extend the road to Peru and pave its entire length. The purpose of the project was to create a more direct route between the lowland rain forest areas of the two countries (CAF 1993). While the project has yet to secure financing, it remains high on the wish list of local authorities, who, instead of waiting for funding, have begun to make piecemeal improvements to the infrastructure on their own.

An improvement on the scale contemplated by the government's original proposal could have severe biodiversity consequences. Land reform is a top policy priority, such that any newly accessible area is sure to be the focus of a major colonization event. In addition to impacting land along its own trajectory—given current pressures in the area for land and access to timber resources—a major project would very probably lead to the ad hoc development of numerous feeder roads, some traversing the Madidi National Park.

In contrast to the Itacare case, when researchers from CI first studied the Madidi situation in 1996, road plans were still at an early and uncertain stage. That allowed them to contemplate influencing the road itself, as well as working for substantive mitigation and/or offsetting conservation investments. Obvious goals included directing any road extensions and improvements away from the most environmentally sensitive areas, avoiding paving (see Pfaff [1997] for discussion of the impact of paving), and in the event a major infrastructure investment is made, negotiating an offsetting investment in the consolidation of the Madidi National Park.

Conservationists laid the groundwork for this strategy with a two-part analysis of the road proposal (see Reid and Landivar [1997]). The first was an economic analysis. The road appeared to be economically questionable because it would have a very small number of beneficiaries (relative to a settled area or the route to a major port) and would require nearly 350 kilometers of pavement to reach Peru. Making the economic argument is important in these circumstances, because decisionmakers who are not sympathetic to nature conservation may view the choice as one between vague environmental values and clear, quantifiable economic benefits. A more potent argument for this audience can be made if it can be shown that scarce public resources are being squandered on a capricious investment with little economic return.

The second part of the analysis examined the biodiversity risk posed by the road. This straight conservation rationale is indispensable for two reasons: First, national policy and lending guidelines of major development banks commonly call for protection of critical habitat threatened by an infrastructure project; second, arguments focused on preserving species often resonate more with national and international publics than dry observations on a project's internal rate of return.

The study found that the project was economically unfeasible and presented very large biodiversity risks. The economic portion of the study employed the same cost-benefit model (the Highway Demand and Maintenance Standards Model) used by road planners at the leading development banks and Bolivia's Servicio Nacional de Caminos. The model results revealed that paving a road to Peru would result in a net present value of negative $25 million. In other words, the cost of building and maintaining the road would be $25 million greater than all the benefits derived by all of its users. This figure translated into a negative 7 percent IRR. At the time of the analysis, development lenders required a positive 12 percent IRR in order to approve a project.

The study found that, relative to other potential projects in the Bolivian Amazon, the biodiversity risks would be considerable. This conclusion followed from an analysis using the Cóndor mapping program to overlay the road on a map of ecoregions derived from E. Dinerstein and others (1995). This exercise showed that the road would affect the Southwest Amazonian and Bolivian Yungas ecoregions. Both of these are classified in the highest category for species richness, with the Yungas system also in the top category for species endemism. The Amazonian portion of the area was in the second highest category for species richness.

Despite these strong results, road development remains a significant threat in the Madidi region. While no international lenders have agreed to finance a large-scale infrastructure project, local pressure for more and better roads is strong. Bridges are being constructed over several of the rivers that up to now could only be forded during low-water periods. One significant feeder road to the existing gravel route is slowly creeping into the hills; people are using fuel and machinery diverted from official maintenance activities to build it. On the drawing board is an alternate route from the highlands that would run through the heart of the Madidi National Park.

While a white elephant highway in the middle of the jungle seems at times a remote possibility, the decentralized nature of infrastructure development in Bolivia makes environmentally and economically senseless investments increasingly likely. Put in national perspective, such northbound routes linking to Bolivia's Amazonian neighbors make little sense. Much more strategic investments can be made linking the country to southern Brazil, Peru's Pacific coast, Argentina, and Paraguay (World Bank 1993). But recent constitutional changes grant a greater measure of authority for infrastructure to provincial governments, whose responsiveness to local political sentiment makes ill-advised projects possible.

Conclusion

Road development threatens many of the remaining undisturbed forest ecosystems in the tropics. But it also presents some of the largest, most promising conservation opportunities. That's because a single decision on a road can have vast conservation consequences—either positive or negative—and the people making those decisions are publicly accountable policymakers, not intractable masses or market forces. Experience is showing that for conservationists to improve environmental outcomes along the lines of the four basic "rules of the road" described earlier, they should follow a few simple strategies, as follows.

- *Start early.* Future road projects must be anticipated before they proceed too far in the planning process to be changed or stopped. Early analysis and intervention allows conservationists to consider the broadest array of options in confronting the threat.
- *Build strong technical arguments.* Roads are among the most popular public investments governments make. Therefore, purely emotional appeals against a road are likely to be met with equally strong emotional appeals in its favor. Technical arguments can augment moral suasion by illustrating the real costs and benefits—quantified and unquantified—that a project entails. Significantly, many roads, especially those in remote areas, are economically unjustifiable but are never criticized on those grounds.
- *Identify common ground with local leaders.* In places where decisions are strongly influenced by local or provincial politicians, road alternatives need to be conceived to overlap with, though not necessarily match, local priorities.
- *Have a plan B.* In some cases, conservationists may be unable to stop or modify a project that threatens to wreak destruction on critical ecosystems. If such a project is sure to proceed unchanged, conservationists should do everything possible to leverage offsetting conservation investments either in the vicinity of the project or, failing that, in another area with high conservation value.

The preponderance of current road investments in the tropics are proceeding without any independent, analytical scrutiny from conservation advocates. The strategies outlined here don't ensure success, but they can greatly increase the chances of it; they deserve increased attention in efforts to conserve the biological riches of tropical forests.

Acknowledgments The author wishes to acknowledge the leading roles played by Rui Rocha, Gerado Bressan, and Gustavo Fonseca in taking the Serra do Conduru State Park from concept to reality. Thanks also to Roger Landivar for his collaboration on the study of roads in Bolivia.

REFERENCES

Bolivia. 1996. "Programa Regional de planificación y Manejo de Areas Protegidas de la Región Amazónica UE-TCA-FAO." Document for binational meeting on the Acuerdo de Cooperación Amazónica Boliviano Peruano.

Chomitz, K. M., and D. A. Gray. 1996. *Roads, Land Use and Deforestation: A Spatial Model Applied to Belize.* Working Paper no. 3. World Bank, Washington, D.C.

Conservation International. 1991. *A Biological Assessment of the Alto Madidi Region and Adjacent Areas of Northwest Bolivia.* RAP Working Paper no. 1. Conservation International, Washington, D.C.

Corporación Andina de Fomento (CAF). 1993. *Road Projects for Integration in the Andes.* Corporación Andina de Fomento, Caracas, Venezuela.

Cropper, M., C. Griffiths, and M. Muthukumara. 1997. *Roads, Population Pressures, and Deforestation in Thailand, 1976–89.* Policy Research Working Paper no. 1726. World Bank, Washington, D.C.

Dinerstein, E., D. M. Olson, D. J. Graham, A. L. Webster, S. A. Primm, M. P. Bookbinder, and G. Ledec. 1995. *A Conservation Assessment of the Terrestrial Ecoregions of Latin America and the Caribbean.* World Bank, Washington, D.C.

Fearnside, P. M. 1987. Deforestation and international development projects in Brazilian Amazonia. *Conservation Biology* 1(3):214–220.

Mamingi, N., K. M. Chomitz, D. A. Gray, and E. Lamblin. 1996. *Spatial Patterns of Deforestation in Cameroon and Zaire.* Working Paper no. 8. World Bank, Washington, D.C.

Pfaff, A. S. P. 1997. *What Drives Deforestation in the Brazilian Amazon? Evidence from Satellite and Socioeconomic Data.* Policy Research Working Paper no. 1772. World Bank, Washington, D.C.

Reid, J. W., and R. Landivar. 1997. *Economic and Biological Consequences of Road Building in the Bolivian Lowlands: A Rapid Assessment Method.* Bolivia Sustainable Forestry Project, U.S. Agency for International Development, Bolfor, Santa Cruz, Bolivia.

Reid, J. W., and R. E. Rice. 1997. Assessing natural forest management as a tool for tropical forest conservation. *Ambio* 26:382–386.

Rice, R. E., R. E. Gullison, and J. W. Reid. 1997. Can sustainable management save tropical forests? *Scientific American* 276(4):34–39.

Sader, S. A. 1995. Spatial characteristics of forest clearing and vegetation regrowth as detected by Landsat thematic mapper imagery. *Photogrammetric Engineering and Remote Sensing* 61(9):1145–1151.

Sader, S. A., T. Severr, J. C. Smoot, and M. Richards. 1994. Forest change estimates for the Northern Petén region of Guatemala—1986–1990. *Human Ecology* 22:317–332.

Thomas, W., W. A. Carvalho, and A. Amorim. 1998. Plant endemism in two forests in Southern Bahia, Brazil. *Biodiversity and Conservation* 7:311–322.

World Bank. 1993. *Bolivia: Transport Sector Strategy.* Report no. 11899-BO. World Bank, Washington, D.C.

18 Cóndor

Better Decisionmaking on Infrastructure Projects

Silvio Olivieri
Claudia Martinez

The Andean Region is an intricate ecological mosaic that possesses the highest biodiversity on the planet. It is rich in hydrological and mineral resources and is home to a great diversity of people with different cultures, different approaches to development, and different ways of relating to nature—from the Peruvian and Bolivian highlands, where indigenous people are living at the same pace as their ancestors, to cities like Bogota, Caracas, or Lima with their large populations, rushing forward into modern life.

In the 1950s, South America accounted for 12.5 percent of the world's trade; by 1990 this percentage had fallen to 3.5 percent. Although several initiatives to promote regional integration and enhance its position in the world economy have taken place in these decades, South America has only recently begun advancing toward a regional development strategy. The strengthening of the Andean Community of Nations, the making of Mercosur, and the idea of establishing a free trade area of the Americas are indications of an increasing emphasis on regional development.

Integration does not only involve trade. It involves common agendas, common frameworks, and a consensus on the long-term future for the region. It is also about physical integration and the necessity to look beyond national boundaries for regional complementarities and global efficiency. Latin America needs a sense of a common future, where the answers to its physical development will be a function not only of geopolitics but of geo-economics and the conservation of the region's rich natural heritage. How much energy is needed to power a country? Could surpluses or deficiencies in energy demands be matched by neighboring countries? How can we make better use of space by building intermodal corridors of integration where roads, electric lines, and gas pipes use the same axis? How can we make better use of water resources by taking transboundary basins into consideration? How can we accomplish better development for the region while conserving our environment and natural resources?

Although South America has shown sustained development during the past decade, it is crucial to understand whether it will continue to be sustainable. Since the 1950s, South America's population has grown to more than 320 million; in the five countries of the Andean Community population has more than tripled, from 30 million in 1950 to about 100 million in 1995. This population increase, compounded by increased per capita consumption levels, has resulted in greater pressure on natural resources and the environment, resulting especially from the opening of their large wilderness areas to colonization. The negative impact of this development on biodiversity and the region's natural capital has been great and has not necessarily resulted in an equally important improvement in the quality of life of its people. So the challenge remains: How can we help the development and integration process become sustainable, better conserve our natural patrimony, and reinforce the regional integration process?

The infrastructure necessary to sustain this development has been controversial from an environmental point of view and is a key issue in the regional integration process. Road-building is a particularly important factor in the rapid and often anarchic colonization of new areas. Such issues underscore the limits of our knowledge of these ecosystems and of our engineering capabilities, when even well-thought-out projects result in large and unforeseen environmental impacts without necessarily achieving their economic or social objectives.

This paper presents a computer-based analytical software tool, called Cóndor, designed by Conservation International (CI) and Corporación Andina de Fomento (CAF). It is meant to be used by everyone involved in developing infrastructure projects and all those interested in the region's sustainable future. Cóndor proposes a new way of thinking, a new approach toward better planning and decisionmaking on infrastructure projects.

The Traditional Approach to Projects and Infrastructure Development

Infrastructure development during past decades was dominated by plans conceived in the 1960s and 1970s around geopolitical theories on how to secure international borders and colonize new territories. Regional integration itself was not necessarily the goal.

As environmental concerns about the opening of new roads have grown, environmental impacts of new projects have been receiving more attention, and most financial agencies now require an environmental impact assessment (EIA) on new projects. Nevertheless, this approach is only a partial solution, necessary but not sufficient to ensure the best decision.

Although policies requiring better EIA studies should be expanded and reinforced in the region, this approach does have limitations, some of them specific to the case of large infrastructure projects.

One such limitation comes from the fact that an EIA is generally done after a project has been designed. This means that in many cases environmental concerns are not addressed in the design phase of the project. This limits our ability to eliminate those alternatives that have high impacts on biodiversity or to build in environmental safeguards. The results of the EIA are of little use in influencing the project, as they arrive late in the decision process and normally lead to a polarization into a "do it" or "scratch it" situation, limiting the consideration of other potentially better alternatives. In most infrastructure projects, this "go/no go" decision tends to be political, made before the technical studies are done. Because an EIA often follows the decision to move ahead with a project, its influence is reduced and in most cases only results in the proposal of mitigation measures.

Another limitation of the traditional EIA approach is that most projects are promoted, designed, and carried out by one ministry within the government, be it the public works, energy, agriculture, or another ministry. Little effort is made to include diverging or conflicting interests of other sectors. For example, while an agriculture ministry might favor colonization of an area, the environment ministry might seek its preservation. Usually one of the competing agencies "wins," often resulting in an inefficient use of resources and social and environmental impacts that are higher than necessary.

If conflicts among sectors within one government are difficult to visualize and solve, a similar situation appears when considering the necessary infrastructure for sustaining regional integration among countries. This effort should be intersectorial and intergovernmental, although most governments are not organized this way. In the end, the traditional way of generating a regional vision has been to sum the parts: the proposals by different sectors and by different countries. This "regional view" is rarely what is needed from the perspective of regional integration and sustainable development.

Improvements made on existing minor roads sometimes result in the construction of integration corridors before a serious study has been done of needs and opportunities that might lead decisionmakers to consider alternative scenarios. A local, unpaved road created by very local demands and pressures can easily end up being part of a major international highway without consideration of its economic, environmental, and social impacts. A major hydroelectric station that operates with net losses and has major social and environmental costs can be built without considering another alternative—importing energy from a neighboring country with electricity surpluses. A gas pipeline can be built just kilometers away from a major highway, opening a parallel access road and destroying acres of forests.

This conventional way of thinking has resulted in inefficient use of resources, slowing of the integration process, and postponement of the urgency of focusing on the sustainability of the process and the conservation of our natural patrimony. It has also left South America with several examples of what not to do; yet the same approach continues to be applied to new projects. After the national level fiscal and monetary readjustments or "first generation" economic reforms, many answers to these errors have been considered, and new adjustments have been made. First of all, the privatization process that is taking over the development of infrastructure is meant to be more efficient. Roads, electric ventures, gas exploitation, and water systems are being privatized throughout the region. Countries are opening their doors to look at their neighbors in search of regional integration projects.

Much analysis has been done on financial efficiencies for all parties involved, but have these new private-sector actors thought in terms of sustainable infrastructure? Are current infrastructure projects better conceived in terms of environmental and social efficiency? Will the planning for new regional infrastructure be structured to address trade agreements or land planning? Are some projects being developed as answers to private pressures and commissions?

Why Cóndor?

Analyzing the way infrastructure is planned and developed, even under new financial and environmental efficiency parameters, CAF, a financial institution that promotes sustainable development and regional integration in Latin America, understood the necessity of promoting the development of sustainable infrastructure. The association with CI, an organization that works for the conservation of the environment, marked the beginning of a new direction in the goal of sustainable development, based on open and honest discussions and objective analysis toward reconciling economic, environmental, and social goals.

After meetings where experts in economic, social, and environmental issues shared their visions, there was a general consensus on the need to create a new computer software tool with the best information available in order to facilitate informed decisionmaking.

Thinking about Infrastructure under Sustainability Criteria in a Regional Context

One of the key issues identified during the initial discussion was the need to facilitate the analysis of infrastructure projects in a regional context that would explicitly highlight the environmental, social, and economic interactions of the project. This would provide the opportunity for every actor along the project cycle to have a regional and integrated view when analyzing the project and to be aware of its consequences to the region's sustainable future. This means that the analytic tool needs to integrate information from a diversity of sectors and should not be limited only to those aspects or areas directly linked to one project. The user should be able to analyze different alternatives to one project on a regional context, to compare one project to another, and to examine overlaps and complementarities.

This ability to examine the regional framework of the project at every step of the project cycle will result in more efficient designs, in the highlighting of complementarities among projects, and in less impact overall on nature and the environment.

After one year of compiling relevant information, a database of more than thirty variables was integrated in a geographical information system. The information compiled was the best available current data. Making this information available to the general public became one of the next goals of the Cóndor project.

Information Access and Use

A common problem in project analysis is the fact that information needed for the analysis is not only dispersed but generally not accessible. Even when an institution does have an information culture, information is managed in a centralized way. The only way to gain access to that information is to address the information sources such as libraries and government agencies directly. In many cases, these institutional sources of data have their own mandates; information demands from external users do not receive a high priority, and response is slow.

Besides, what is required is the combined expertise in information management and specific technical, analytical methods, which will vary from decisionmaker to decisionmaker. The compilation of data is often driven by a given project manager's or counsultant's experience or set of contacts. Creation of a broad and unbiased dataset requires a greater investment in a full information search. It is very difficult for each analyst to have knowledge of the environmental, social, or economic factors that might affect the project outside his or her own field of expertise.

What is needed is a process where information is openly available for the technical analyst, or for any interested party, to do his or her own analysis, using his or her own criteria and methods and directly manipulating the relevant data.

In this sense, the use of a Geographical Information System (GIS) is just the first step in developing a new tool to help decisionmakers think differently. The second goal becomes the need to produce a tool for nonexperts that makes them see the different aspects involved in the development of infrastructure projects and, more important, gives them a sense of regional development in a broader context than can be gained on just a project-by-project basis.

A New Way to Make Decisions

As just described, information being generated is not being used on a regular basis, and if people want to use it, they face lengthy and bureaucratic procedures. As information is not applied to meet actual professional needs, its quality is seldom assessed.

Our experience with current information use shows access to useful data often requires a radically different approach based on bypassing the intermediary—that is, the information units—and on easy access to information for a wider diversity of users. In this context, information managers should perform less "custom-made" analysis and orient their institutions to the development of information products and tools that anyone can use and to the tasks of keeping the tools and the underlying information current. These tools should be widely available to all stakeholders.

In order for this tool approach to succeed, the tools should be carefully designed to integrate the appropriate mix of information at the right scale, resolution, and quality. The tool should also be conceived and developed as a "zero-learning" tool, meaning one that does not require additional training to use it, and it should involve a soft learning curve that would ensure its widespread use.

This approach would ensure a shared and common basis of information to understand the regional context of infrastructure for every stakeholder in the project cycle. This shared information will result in better projects and better integration in the regional context.

This would be a tool that helps give users a broad perspective of available options before the projects are already decided, a tool that can be used to alert decisionmakers to the tradeoffs of building this or that project. It is also a way of introducing checks and balances into the early conception of any project.

Most important, this would be a tool that gives users a vision for Latin America—a vision of future populated areas, of pristine environments and their importance for the future, of the industrial poles,

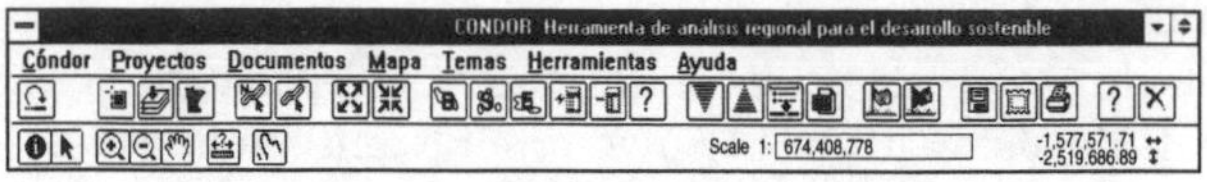

Figure 18.1 Cóndor tool bar.

of deforested and degraded areas, and of all the opportunities we have in our hands to shape Latin America, and the challenges they bring.

Cóndor is meant to fly into the future and show us Latin America in fifty, one hundred, or two hundred years.

How Cóndor Works

Cóndor, the bird, flies high and has a sharp eye. It flies high enough to see the broad picture as well as the details. Cóndor, the tool, tries to implement the same principle by facilitating the evaluation of a road or other infrastructure project in a regional context and spotting potential areas of conflict between the project and its regional environment as expressed in biodiversity, social, and economic factors. It is a tool intended for all stakeholders along a project cycle, from concept to design to implementation. It is intended for the broad, early evaluation of project ideas or proposals and for the analysis of alternative designs using a familiar mapping interface. It is also intended as a zero-learning tool: an easy-to-use tool that any person familiar with Windows should be able to use without special training.

Cóndor consists of both a geographic database of the region and an easy-to-use, targeted analysis program. The capabilities of Cóndor range from simply browsing the maps and visually realizing the correlation, or lack thereof, between themes in a project area, to composing semidetailed maps of a project and printing or exporting these maps for further analysis (figure 18.1).

As an exploratory tool, Cóndor allows users to focus on the region of concern by overlaying different geographic layers representing environmental or socioeconomic variables. He or she can visualize the interactions among specific variables and indices, identify potential concerns, and focus on critical areas of a project.

In this process, the user can start by exploring one of the roads included in the database of 170 existing and proposed road projects in the region to analyze its potential impacts (figure 18.2). He or she can also generate new road projects, or different scenarios or variations of one project, in order to analyze and compare the potential benefits or problems implied. The user would do this by digitizing a new road or a modified design directly on screen with the help of the mouse, using existing roads, rivers, and other geographic features as help.

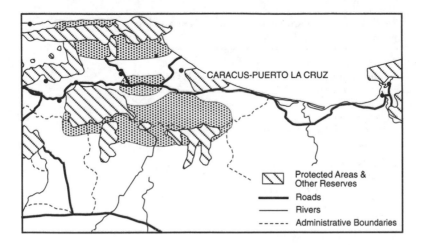

Figure 18.2 Caracus-Puerto La Cruz.

As an analytical tool, Cóndor can perform an "environmental warning" or flagging analysis, which consists of establishing an area of influence for a road project defined as a buffer or corridor along the road, of a width defined by the user. Cóndor will then check key variables and their values within this area of influence and flag the ones that reflect potential problems for that road project. The result is a written report (table 18.1) and a map that highlights these potential concerns. Analysis may show that the road project crosses an area of great biological importance or that it falls near or crosses an indigenous reserve, a national park, or an area of steep slopes. The user can then modify the road design, analyze alternative pathway options, and compare their performance.

The problems flagged by Cóndor are potential conflicts or impacts. The user is still responsible for seeking expert help in understanding how these potential problems affect the project or how the project proposal is addressing them. In this way, Cóndor facilitates the development of better terms of reference for projects.

As a reporting tool, Cóndor facilitates the composition of maps of any road project or area within the Andean Region, selecting the specific area and the mix and appearance of variables and indices. These maps can be easily printed in standard or large-format printers, or exported as graphics that can be easily integrated into word-processing documents.

Finally, Cóndor also gives access to reference documents contained in the database that can help the user better understand the environmental concerns identified by the analysis. These documents, generated along the course of the project, include analysis of population tendencies in the region, guidelines for road design, and so on as well

Table 18.1 Part of an Analysis Report Generated by Cóndor on a Previous Road Project*

Reporte de Análisis
Nombre del proyecto: CARACAS - PUERTO LA CRUZ
Tipo de proyecto: CARACAS - PUERTO LA CRUZ

Areas intersectadas
Las áreas intersectadas son áreas geográficas que son cruzadas por el proyecto de infraestructura analisado.

Factores sociales (importancia):
Nombre del área: Cordillera Caribe
Importancia factores sociales: 4
Nombre del área: Anzoategui
Importancia factores sociales: 3

Venezuela: Areas reservadas:
Nombre del área: El Avila NP
Tipo de área: Parque Nacional
Nombre del área: Alta y Media Guajira Resg
Tipo de área: Resguardo Indigena
Nombre del área: Laguna de Tacarigua NP
Tipo de área: Parque Nacional

Importancia biológica por ecoregión:
Nombre de la ecoregión: Bosques Montanos Cordillera La Costa - Venezuela
Descripción: muy alto impacto por actividades humanas, importantes parques nacionales; bosque de neblina muy rico en epifitas.
Nombre de la ecoregión: Bosques Montanos Cordillera La Costa - Venezuela
Descripción: muy alto impacto por actividades humanas, importantes parques nacionales; bosque de neblina muy rico en epifitas.
Nombre de la ecoregión: Bosques Montanos Cordillera La Costa - Venezuela
Descripción: muy alto impacto por actividades humanas, importantes parques nacionales; bosque de neblina muy rico en epifitas.

Clases de pendientes:
Tipo de suelo: Orthic_Acrisols
% de pendiente: 30%
Clase de pendiente: 5
Tipo de suelo: Luvic_Xerosols
% de pendiente: 30%
Clase de pendiente: 5

Areas al interior del corredor

Venezuela: Areas reservadas:
Nombre del área: Mochima NP
Tipo de área: Parque Nacional

Clases de pendientes:
Tipo de suelo: Luvic_Xerosols
% de pendiente: 30%
Clase de pendiente: 5

*The report is a text file in Web/HTML format.

Table 18.2 Contents of the Cóndor Database (Map Layers)

Base information
  International limits
  Provincial, departmental or state limits
  Cities and towns
  Existing roads
  Integration corridors
  Proposed roads

Natural Environment
  Topography
  Hydrology
  Climate (annual temperature and rainfall)
  Soils

Biodiversity
  Ecoregions
  Biologically important areas
  Species diversity, endemism, higher level diversity and endemism
  Habitat diversity and fragility
  Forest cover

Socioeconomic
  Protected areas
  Areas of social concern
  Indigenous reserves and other forms of land reserves
  Population census data for 1980, 1990, and projection to 2000 (population,
    population growth, and density)

as additional references the user can use to gain further understanding of the problems.

The geographic database at the core of Cóndor covers the five northern countries of the Andean Region, at a scale of one to one million,
with more than thirty variables related to the natural environment,
biodiversity, social, and economic issues. (A detailed list of information and data layers contained in Cóndor is presented in table 18.2).

The catalog of road projects contains about 170 projects for the development or improvement of roads proposed by the governments of
the region as part of existing or planned integration corridors.

Future Development of Cóndor

Many people in the region have already used Cóndor 1.0. Six months
after the presentation of Cóndor in the Andean Region, our early assessment of its use has been very positive. First, Cóndor is successful
because of the amount of information that it makes available at many
levels within institutions. This information, when linked to the analysis of or decisionmaking on road projects, gives direct control to the

user. Cóndor users are less dependent on results or inputs from their information centers or specialized offices, and they achieve greater efficiency. Second, Cóndor is easy to use. Little or no training is necessary, so it can be used by professionals of very different backgrounds (engineers, financial officers, policy personnel, etc.).

Some of the users include the loan officers of AFC, who have Cóndor on their computers as a daily tool to help them analyze the feasibility of any project from its conception. The experience shows that Cóndor helps develop better project analysis. The interaction with all variables involved in infrastructure development makes decisionmakers define more comprehensive terms of reference from the beginning of a project. Ministries like the Environment Ministry of Colombia want the tool to be expanded and used by all their personnel. Several private companies have also asked for the tool because of their daily need to have access to better information.

During the Second Iberoamerican Workshop of the Environmental Units of the Transport Sector, held in Cartagena, Colombia, in February 1998, training on Cóndor was provided to personnel of different government agencies of the region. Their feedback in the use of this tool to facilitate broader perspectives about infrastructure development choices with colleagues was excellent.

Based on the feedback received from the users, a second version, with an improved interface and functionality and an updated database, was designed and made available in mid-1998.

What's next? First, there is a need to keep the Cóndor database up to date, which means that a new version of it should be produced yearly or every two years. In the meantime, some of the more dynamic aspects of the database can be made available and updated through the Internet, and a technical support Web page is kept current. This information update can easily be done with the collaboration of the users themselves, including the key institutions responsible for infrastructure projects. In fact, as their information is being used by a larger number of people for a diversity of analysis, new demands and feedback are being generated that will motivate better quality and quantity of information, as well as promote better analysis.

Our evaluation of the tool concept represented by Cóndor and its impact in the region has generated interest in expanding the concept and developing a more powerful tool. We are currently working on a new version with larger geographic coverage, a larger database with more information layers and indices, with an enhanced balance of themes, including new variables in all categories, including natural environment, biodiversity, economic issues, social aspects, and so on. This version also focuses on different kinds of infrastructure projects, including electric lines, gas pipes, irrigation projects, and so forth.

Conclusion

Cóndor integrates the best regional information on environmental, social, and economic factors with the best information on proposed or active projects, within an easy-to-use framework. Its capacities for exploring and analyzing integrated information facilitate the development of better projects with better insertion in a regional context while paying attention to their sustainability and their impacts on biodiversity and the environment.

The tool is based on a new way of approaching infrastructure project design and evaluation. This approach is based on more open access to information, the sharing of a common information base, and open discussion of a project's positive and negative aspects and alternative options.

By providing more information and making it accessible in an intelligent way, Cóndor ensures better decisionmaking and less conflict. By introducing the regional view in the project analysis, it establishes a better way of developing a common view for the region's future. By favoring widespread availability of information, it will develop a demand for more relevant and better quality information.

As a pilot project, still under evaluation, we expect Cóndor to prove that this new way of thinking about projects and regional development is feasible and interesting. We also hope that Cóndor will help the development of new and better tools based on this approach to help us in deciding the kind of future we want for our region.

REFERENCES

Batista da Silva, Eliezer. 1996. Infrastructure for Sustainable Development and Integration of South America. Caracas, Venezuela-Business Council for Sustainable Development-Latin America and the Andean Development Corporation (BCSD-LA/CAF).
CAF and CI. 1997. *Cóndor: Herramienta de Análisis Regional para el Desarrollo Sostenible, Version 1.0.* Caracas, Venezuela and Washington, D.C.

PART VI

Conclusion

Conclusion

Leave More Than Footprints

The New Corporate Responsibility

Glenn T. Prickett
Ian A. Bowles

American lovers of the outdoors have a slogan: "Take only photographs, leave only footprints." As is shown in this book, resource extraction in rich tropical ecosystems—the ultimate "outdoors"—takes far more than photographs and can leave devastating footprints. In some cases, like national parks, other protected areas, and the lands of indigenous peoples who oppose extraction, this means that natural resources should stay in the ground. Where development proceeds, however, companies have an obligation to minimize their impacts and to leave more than small footprints; they should also contribute to biodiversity conservation and the welfare of local people.

Conservationists see this as a moral obligation. A private company that profits from an ecosystem's resources should make the investments required to ensure that the land's other values—from local subsistence needs to global biodiversity benefits—are maintained. Governments are beginning to enact this obligation through laws and policies that require environmental impact assessments (EIAs) and social impact assessments (SIAs), best management practices, and revenue sharing with local communities and conservation areas.

Enlightened companies, meanwhile, are beginning to see conservation and community development initiatives as good business. Often this extends beyond compliance with government requirements. Voluntary action to minimize environmental impact and to advance conservation and community development can enhance a company's reputation in the eyes of regulators, consumers, shareholders, employees, and citizen activists, whose opinions ultimately affect a company's financial performance. Voluntary investments can also help reduce the likelihood of costly delays due to government and public criticism and expensive remedial measures that might be required were environmental and social damage to occur. In some cases, environmental investments can even lower costs.

The case studies in this book describe the motives, actions, and initial positive results of companies who recognize environmental and social responsibility as a "cost of doing business." The overview chapters on oil and gas, mining, timber, and infrastructure, on the other hand, depict the damage done by companies with weak environmental and social standards.

Given these risks, why not place all sensitive ecosystems off limits to resource extraction? Such a conservation strategy, while appealing on the surface, is unlikely to succeed. The biodiversity hotspots and tropical wilderness areas described by Russ Mittermeier and Bill Konstant in chapter 1 are already pockmarked with oil and gas, mining, and timber concessions granted by governments to resource companies. Figure 1 graphically depicts this phenomenon in the tropics of Latin America. While some of these projects might be successfully challenged, developing country governments will certainly be unwilling to forego all of the revenue they can earn through resource extraction.

A balance must be struck between areas that should be placed off limits—national parks, other protected areas, and lands of indigenous peoples who oppose resource extraction—and other areas where resource extraction can proceed. Where development happens, we conclude from cases presented in this book that a company's own commitment to "do it right" is essential if a resource development project is to have acceptable impacts on sensitive ecosystems and local communities. Government regulation and citizen activism are critical, but by themselves they are often ineffective in the remote parts of the developing world that form the new frontier for natural resource extraction.

This statement is not a naive appeal to corporate altruism. Companies will act in their own interests. If higher profits can be made by "cutting and running"—too often the case in the past—then environmental and social damage will result from resource extraction. The cases of corporate responsibility described in this book resulted not from altruism but from a business environment in which companies had clear incentives to "do the right thing."

Business Incentives for Environmental and Social Investments

On the surface, this would appear to be a paradox. The opening of natural resource industries in developing countries to international

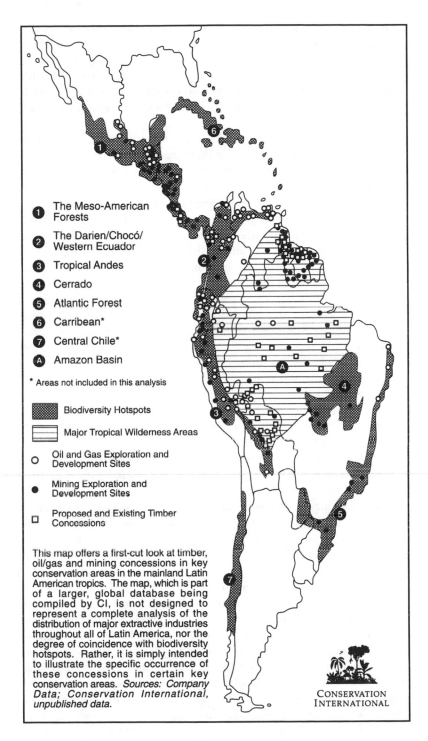

The following text appears within the map figure:

1 The Meso-American Forests

2 The Darien/Chocó/ Western Ecuador

3 Tropical Andes

4 Cerrado

5 Atlantic Forest

6 Carribean*

7 Central Chile*

A Amazon Basin

* Areas not included in this analysis

Biodiversity Hotspots

Major Tropical Wilderness Areas

○ Oil and Gas Exploration and Development Sites

● Mining Exploration and Development Sites

□ Proposed and Existing Timber Concessions

This map offers a first-cut look at timber, oil/gas and mining concessions in key conservation areas in the mainland Latin American tropics. The map, which is part of a larger, global database being compiled by CI, is not designed to represent a complete analysis of the distribution of major extractive industries throughout all of Latin America, nor the degree of coincidence with biodiversity hotspots. Rather, it is simply intended to illustrate the specific occurrence of these concessions in certain key conservation areas. *Sources: Company Data; Conservation International, unpublished data.*

CONSERVATION INTERNATIONAL

Figure 1. Extractive concessions in mainland Latin America tropics.

competition puts pressure on companies to cut costs. Voluntary spending on environmental and social measures would seem to fly in the face of companies' business incentives in an era of globalization. Yet this book contains examples of some of the world's most profitable resource companies going "beyond compliance" to address environmental and social issues. Why? Our case studies suggest two key incentives.

The first is reputation. At a global level, natural resource industries are under fire as never before for their legacy of environmental and social harm. The Bank Information Center's Lisa Jordan and Christopher Chamberlain, reviewing a series of controversial projects in chapter 4, conclude that "mining, logging, and petroleum companies seem to be emerging in the public eye as the new 'pariah' industries." Their conclusion is supported—in the United States and Europe at least—by a steady stream of shareholder resolutions, full-page newspaper ads, and public demonstrations against resource companies and their operations in the tropics.

These attacks on corporate reputations can exact real—if hard to quantify—costs in companies' operations. In the field, protest and controversy can delay project implementation schedules and force costly changes in project design. In some cases, companies may be forced to abandon projects altogether after months or years of preparatory investment. Upon earning a negative reputation on environmental and social issues, companies may face tougher negotiations on new concessions. At a global level, controversy can harm the brand image of companies that deal directly with consumers, especially in Europe and North America. Finally, controversy can affect employee morale and make it harder for companies to attract and retain young workers—who may be members of the environmental groups leading the antibusiness campaigns.

The value of corporate reputation and its vulnerability to criticism on environmental and social grounds was the feature of an unusual 1997 report by the Control Risks Group, entitled "No Hiding Place: Business and the Politics of Pressure" (Control Risks Group 1997). The report was unusual not for its subject matter—growing public concern about the environmental and social impacts of resource development—but for its author, a business consulting firm. The report concluded that worldwide citizen activism, well organized and electronically networked, has left "no hiding place" for irresponsible resource development. To gain a "license to operate," companies must adopt the highest environmental and social standards. "No Hiding Place" demonstrates that globalization creates new constraints for resource companies as well as new opportunities. In the age of the Internet, activists can scrutinize far-flung operations as closely as those close to home—with equally strong implications for regulatory approvals, consumer acceptance, shareholder value, and employee loyalty.

As demonstrated in this book, leading companies are responding to heightened public scrutiny by devoting additional resources and management attention to environmental and social issues. Some are going beyond a "do no harm" approach and actively supporting conservation and community development programs.

In addition to corporate reputation, a second incentive driving resource companies to invest in environmental measures might be called enlightened economics. In industry in general, many case studies have shown that environmental investments can improve profitability, especially through "ecoefficiency"—using fewer raw materials and generating less waste (DeSimone and Popoff 1997). Ecoefficiency measures have helped firms reduce operating costs, avoid costly environmental mitigation measures and liabilities, and boost sales through new and differentiated products.

In principle, such gains should be possible in resource extraction projects. While little analysis exists of the costs and benefits of environmental practices in natural resource operations in the tropics, several examples suggest that "win-win" approaches may be possible. Directional drilling in oil development—running multiple wells from a single well pad—can minimize both the footprint and the costs of an operation. Case studies from Block 16 of Ecuador's Oriente (Amazon) region and the Ras El Ush oil field in Egypt's Gulf of Suez confirm that directional drilling can save companies millions of dollars and reduce impacts on sensitive forest and marine ecosystems (Vera et al. 1998; Brown and Nour 1998). Avoiding construction of access roads in remote rain forest areas reduces the threat of deforestation and can be cost-effective over the life of a project. ARCO's Robert Kratsas and Jennifer Parnell conclude in chapter 5 that helicopter and monorail transport at the company's Villano field in Ecuador's Oriente will be no more expensive than an access road when twenty years of road maintenance and other expenses are taken into account.

A recent benchmarking survey by Arthur D. Little for a major oil and gas company found that 90 percent of industry respondents believe that environmental performance affects their company's competitive position (Slocum and Willson 1998). The most common explanation was that good environmental performance helps cut costs. Survey respondents also concluded that good environmental performance enhances reputation and eases market entry. All surveyed said that they go "beyond compliance" on environment, health, and safety issues to secure long-term cost savings, demonstrate their commitment to social responsibility, boost satisfaction of employees and stockholders, and win the favor of regulators.

More research is needed on the economics of environmental management in the natural resource industries to determine whether a green invisible hand may be at work. But interesting evidence is emerging from equities markets that strong environmental performance may be a signal of strong financial performance. The Store-brand Scudder Environmental Value Fund, operating in Europe since 1996 and screening investments on financial and environmental criteria, has consistently outperformed the Morgan Stanley Capital International World Index. In the United States, a phantom portfolio of companies with strong environmental records outperformed the Standard and Poor 500 index by 2.8 percent in the first half of 1998 (Deutsch 1998).

If this trend holds for the natural resource industries, globalization may raise the environmental bar for resource extraction projects, not lower it as many activists fear. But concerned citizens and governments should not depend on a green invisible hand. Two of the leading incentives for good environmental behavior—enhancing corporate reputation and easing market entry—depend more on green politics than on green eyeshades.

As regulators and activists debate resource extraction projects in sensitive ecosystems, they should examine companies' environmental and social records, reward only the best with natural resource concessions, and work to hold companies accountable for their green commitments.

Pick Winners

If environmental and human rights activists have accumulated the power to influence the fate of resource development projects, how should they use it? It seems an unlikely question. The image of activists—to the world and to themselves—is that of a humble David confronting a corporate Goliath. No one expects the activists to win.

But win they have. Consider two unlikely examples:

- In March 2000, Mitsubishi Corporation and the government of Mexico abandoned plans to build a $100 million salt plant on the shore of Baha's Laguna San Ignacio after a public campaign led by Mexican and U.S. environmentalists generated close to a million letters, emails, and petitions in opposition to the project. Environmentalists were concerned that the proposed salt works would degrade the last undisturbed birthing and nursery grounds of the gray whale and cause other ecological harm in an area designated as both a World Heritage Site and a U.N. Biosphere Reserve (Preston, 2000).
- In November 1994, the government of Quebec cancelled electric utility Hydro-Quebec's plans for a 3,000 megawatt hydroelectric project that would have involved the construction of dams, dikes, and an artificial lake on the Great Whale River. Local Cree Indians joined with environmentalists to mount an

international campaign against the project, which they charged would flood over a thousand square miles of wilderness, including important Cree hunting grounds (Davis, 1994).

We believe that this power can be applied where resource development does proceed to reward companies who demonstrate strong environmental and social commitments and reject companies who lack them. If, as we believe, a company's own commitment makes a key difference in the environmental and social outcome of a project, activists and governments should make careful assessments of companies' past records and current commitments. They should intervene to support companies with a proven commitment to "do it right" and oppose companies whose environmental and social credentials are lacking or questionable.

Such judgments in practice will never be simple. A company's record and pledges will usually appear in shades of gray, not stark black or white. Yet making the judgments will put activists in a stronger position to influence development where it proceeds and to stop it where it shouldn't. Nongovernmental organizations (NGOs) should develop a better basis for deciding which projects and companies may offer an acceptable outcome—or even improvements—for ecosystems and local people, and which deserve to be opposed.

Making better judgments on resource development projects will require NGOs and activists to improve their knowledge of the extractive industries and to develop the capacity to conduct economic analyses of proposed development projects. NGOs could make a useful contribution by seeking to develop common metrics—perhaps a "footprint index"—with which to gauge proposed resource development projects and to monitor their performance over time. Such an index could address:

- *The affected ecosystem.* Where will the proposed development take place? In what type of habitat? Is it a priority area for biodiversity conservation? Does it enjoy protected status, or has protection been proposed? What are the other threats to ecosystem's biodiversity? Would the proposed project exacerbate them? Or could it counter them?
- *The affected people.* What local communities will be affected and how? Do they include voluntarily isolated indigenous groups? What are local views on the project? Are communities' legitimate rights to land and resources recognized? Does the project conflict with these rights?
- *The company's proposal.* Does the company intend to apply best management practices throughout the life of its project? Has it made adequate plans for public participation? Has it pledged to share benefits with local communities? Will it create or support conservation areas?

- *The company's record.* What has the company's experience been in other, similar projects? How does it rank on general environmental and social indices? Does it have environmental and social advocates on its board of directors or advisory committees? Has it implemented successful partnerships with environmental groups and affected communities?
- *The project's results.* Has the company implemented its environmental and social commitments? Have they worked as planned? How has the project affected the welfare of communities and the state of the ecosystem?

Industry should welcome an objective index or set of measures that would allow the best companies to demonstrate their credentials to skeptical regulators, community members, and activists. The development and application of such measures would stimulate companies to do what they do best—compete—to the ultimate benefit of the environment and local people.

Nor should such an effort be confined to private action. In addition to setting environmental and social standards for all to follow, governments should establish clear criteria for weighing environmental and social commitments and past performance when awarding resource concessions.

Creating objective environmental and social criteria for resource development projects—and systems that allow for the ranking of competing companies—presents a clear and compelling opportunity for partnerships among NGOs, industry, and government. Together we can develop new rules of the game to make the footprints of resource development bearable—and perhaps beneficial—for the world's biodiversity and all those who benefit from it.

REFERENCES

Brown, T. G., and M. H. Nour. 1998. Harmonizing environmental and economic profiles in a marginal oil field development project: A case study. Paper presented at the Society of Petroleum Engineers International Conference on Health, Safety, and Environment in Oil and Gas Exploration and Production, June 7–10, Caracas, Venezuela.

Control Risks Group. 1997. *No Hiding Place: Business and the Politics of Pressure.* Control Risks Group, London.

Davis, Phillip. 1994. Quebec's New Government Cancels Hydroelectric Plant. *Morning Edition*, National Public Radio, November 23, Transcript #1483-2.

DeSimone, Livio D., and Frank Popoff. 1997. *Eco-Efficiency: The Business Link to Sustainable Development.* MIT Press, Cambridge, Mass.

Deutsch, Claudia H. 1998. For Wall Street, increasing evidence that green begets green. *New York Times*, July 19, p. BU-7.

Preston, Julia. 2000. In Mexico, Nature Lovers Merit a Kiss From a Whale. *New York Times*, March 5, p. A-13.

Slocum, Dean A., and John S. Willson. 1998. Using EHS to add business value in international petroleum industries: Some current benchmarking results. Paper presented at the Society of Petroleum Engineers International Conference on Health, Safety, and Environment in Oil and Gas Exploration and Production, June 7–10, Caracas, Venezuela.

Fulleda, Federico. 1998. Occidental, BP to stay off Indian lands. United Press International, May 26.

Vera, Juan Sebastion, et al. 1998. The use of directional drilling in the industry's effort to minimize environmental impact: Block 16 project. Paper presented at the Society of Petroleum Engineers International Conference on Health, Safety, and Environment in Oil and Gas Exploration and Production, June 7–10, Caracas, Venezuela.

Index

land holdings, 149
political power, 236
property rights, 128, 142, 149
seedling supply, 140, 142
Local markets, 13–14, 157
Logging. *See* Timber extraction
Long-term impact assessment, 252
Los Rojas project, 216–217
Lovejoy, Thomas, 225–226
Lower impact forest management, 170

MacKinnon, John, 259
Madagascar, 190
Madidi region, Bolivia, 287–289
Mahogany, 153–154
Major tropical wilderness areas, 20–21
Malaysia, 41, 118–121
Mangrove forests, 78
Marine life, 78
Market share, logging, 119
Marketing certified forest products, 154–155, 157, 160–162
Martin Guitars, 161
Maya Biosphere Reserve, Guatemala, 254
Measuring biodiversity, 10–15, 95, 99–100
Medical services accessibility, 264
Medium and small-scale impacts, 95, 98
Megadiversity countries, 21–24, 147, 186
Megavertebrates, 25
Mercury, 39, 217
Mexico
biodiversity, 147
changes in forestry, 145
community-based forest management, 149
current forest conditions and trends, 145–150
deforestation, 147–149
financing of certified forest production in, 160–164
forest coverage, 146
forest law, 158
FSC-endorsed certification in, 151–157

land rights and ownership, 149
market incentives for improved forest management, 160–162
market share for certified forest products, 160
policy incentives for improved forest management, 158–160
Sociedad de Productores Forestales Ejidales de Quintana Roo (SPFEQR), 152–155
temperate forests, 146
tropical forests, 146
Union de Comunidades Forestales Zapoteco-chinanteca (UZACHI), 156–157
wood products industry, 149
Michael Elkan Studio, 161
Minas Gerais, 140
Mine closure, 192–193
Mine Environmental Neutral Drainage (MEND) program, 213–214
Minera Las Cristinas, 216–217
Minera Penmont, 214–216
Minerals Council of Australia, 212
Minerals extraction, 39. *See also* Nonferrous metals extraction
adaptive conservation management strategies, 188–189
best practices in, 189–193
changing corporate practices, 195–198
cleanup technologies, 190–192
competitiveness, 198–199
concession contracts, 233–244
conservation innovation, 199
dynamic firms, 199
ecosystem effects of, 184–185
environmental monitoring, 242–243
exploration, 189–190
infrastructure, 187
legislation, 193–195, 233–234
Madagascar, 190
megadiversity countries as emerging markets, 186
mine closure, 192–193

Rainforest Action Network (RAN), 41–43
Ramsar sites, 223–224
Ranking timber extraction conservation options, 174t
Rapid Assessment Program, 20, 99
Rapid timber harvest, 171
Reclamation, 63
Recreation, 12–13
Recycling, 206, 210–211
Reduced impact logging techniques, 172, 176
Reference documents, 299–300
Regeneration. *See* Restoration
Regional development strategies, 20, 69, 292–303
Regional markets, 13
Regulatory mechanisms, 35–36, 48, 275
Rehabilitation. *See* Restoration
Repressive governments, 48–49
Reputational leverage, 44
Research, scientific, 13
Reserves. *See* Protected areas
Resilience, environmental, 186
Resource Conservation and Recovery Act (RCRA), 191
Restoration
  disturbed areas, 84, 92
  drop zones, 103
  forest, 156, 171
  helipads, 102–103
  mineral sites, 186, 192
Rio Tinto, 40
Road development and improvement, 61, 83, 86–87
  assessing socioeconomic/development conflicts, 271
  avoiding vulnerable ecosystems, 283
  benefits of, 262–264
  biodiversity risk, 288–289
  black-market activities as a result of, 267
  Brazil, 285–287
  conservation strategies for, 290
  cost-benefit analysis, 264–268, 282–284, 288–289
  deforestation as a result of, 281–282
  drill, 224

effects of weak environmental regulatory agencies, 270–271
  encroachment as a result of, 266–267
  environmental impact assessments, 268–269, 272–273
  environmental technical assistance, '273
  expandability, 282
  GIS as tool in, 276–277
  GMS Action Plan, 278
  impact on indigenous peoples, 267
  improving conservation outcomes of, 283–285
  indirect impacts of, 284–285
  integrated regional planning, 268, 271–272
  international policy research institutions, 276
  irreversible impacts of, 284–285
  Itacare, Brazil case study, 285–287
  land use and, 277
  mineral development, 187
  minimizing adverse impact of private sector, 275
  multiple purposes of, 282
  negative social impact of, 264–267
  paving, 282, 284, 288
  potential negative impacts of, 264–268
  projects in Condor database, 301
  protected areas and, 285–287
  public participation in, 274–275
  scaling quality of, 284
  social impact assessments, 268–269
  socioeconomic initiatives, 262–264, 272
  tapping institutional knowledge, 276
  tenure policies, 277
  Transamazonia Highway, 281–282
  upgrading existing roads, 261–262
  watershed management, 271–272